PRAISE FOR BERND SCHONER AND
THE TECH ENTREPRENEUR'S SURVIVAL GUIDE

"In classic Media Lab 'Practice over Theory' style, Bernd provides a very practical and useful guide to technology entrepreneurs based on experience. A must-read for any tech entrepreneur trying to build a company."

—JOICHI ITO, director of the Massachusetts
Institute of Technology Media Lab

"Bernd is an insightful entrepreneur behind one of the great Kendall Square startups. I followed his story from beginning to end from the outside, while it was actually happening. What a delight to be able to hear the full story from the inside.

"A must for any budding entrepreneur. It explains every aspect of the startup process, from knowing when you've got the right idea to surviving within a large corporation after you've sold your startup—and everything in between."

—TIM ROWE, founder and CEO of the
Cambridge Innovations Center

"With humor and the wisdom of experience, Bernd brilliantly captures the essentials of building, funding, and exiting a tech company. Whatever the problem, *The Tech Entrepreneur's Survival Guide* offers a straightforward and lucid solution to the startup dilemma."

—HANS-MICHAEL HAUSER, managing director at
the Boston Consulting Group

"A crucial and wise guide for anyone considering a journey through startup land, with clear lessons for the budding entrepreneur."

—SANJAY SARMA, professor of mechanical
engineering and director of digital learning
at the Massachusetts Institute of Technology

THE
TECH
ENTREPRENEUR'S
SURVIVAL
GUIDE

How to Bootstrap Your Startup, Lead Through Tough Times, and Cash in for Success

Bernd Schoner, PhD

Mc
Graw
Hill
Education

NEW YORK   CHICAGO   SAN FRANCISCO
ATHENS   LONDON   MADRID
MEXICO CITY   MILAN   NEW DELHI
SINGAPORE   SYDNEY   TORONTO

1 2 3 4 5 6 7 8 9 0 DOC/DOC 1 2 0 9 8 7 6 5 4

ISBN 978-0-07-182397-5
MHID   0-07-182397-2

e-ISBN 978-0-07-182336-4
e-MHID   0-07-182336-0

Design by Lee Fukui and Mauna Eichner

This publication is designed to provide accurate and authoritative information in regard to the subject matter covered. It is sold with the understanding that neither the author nor the publisher is engaged in rendering legal, accounting, securities trading, or other professional services. If legal advice or other expert assistance is required, the services of a competent professional person should be sought.
    —From a Declaration of Principles Jointly Adopted by a Committee of the
        American Bar Association and a Committee of Publishers and Associations

Library of Congress Cataloging-in-Publication Data

Schoner, Bernd.
  The tech entrepreneur's survival guide : how to bootstrap your startup, lead through tough times, and cash in for success / by Bernd Schoner.
    pages cm
  ISBN 978-0-07-182397-5 (hardback) — ISBN 0-07-182397-2 (hardback)
  1. High technology industries—Management. 2. Technological innovations—Management. 3. New business enterprises—Management. 4. Entrepreneurship. I. Title.
  HD62.37.S36 2014
  658—dc23

                                                            2014005514

To Mary
for sharing in the world's greatest entrepreneurial adventure:
building a life and cofounding a family

Contents

Acknowledgments

Many have helped build ThingMagic, and many have helped me write this book. Thank you, all. I am forever grateful for your contributions, dedication, sacrifices, and humor.

Working with my ThingMagic cofounders, Ravi Pappu, Matt Reynolds, Rehmi Post, and Yael Maguire, at MIT and beyond has been the most formative experience in my life. Thank you for being smart, creative, ambitious, and—sometimes—patient!

Only Ravi and I remained employed at ThingMagic at the time of the acquisition. Some 10 years earlier, we were the first full-time ThingMagic employees. We practically sat on each other's laps in our minuscule office, seemingly disagreeing on everything from the room temperature to the prospects of a career outside science. Somehow, we managed to translate those debates into a great working relationship that continues to this day.

Tom Grant, ThingMagic's CEO, began working with us shortly after we incorporated the company, and ever since, he has gently and respectfully steered our group of headstrong founders who didn't really know what was good for them. As a former venture capitalist, Tom knows the two sides of the financing and M&A processes better than anyone, and he brilliantly put his wisdom and experience to work at ThingMagic.

ThingMagic's investor group and board representatives were a lot more patient and supportive than we deserved. We may not have shown our gratitude as much as we should have, but we did appreciate your hanging in there: Marcelo Chao, Larry Begley, Stephanie O'Brien, Arnold Chang, Jeff Williams, and Nicholas Negroponte.

Steve Berglund, CEO of Trimble, and the Trimble family provided ThingMagic with a stable, yet innovative, home. Steve had the great vision to add ThingMagic's technology to the Trimble portfolio as a foundation for new and exciting applications.

Our investment banker, Chris Pasko at Blackstone, did not relent until ThingMagic's acquisition closed. Thank you for making this deal work. It must have been one of the more difficult ones in your career.

Jon Gworek and his team at the law firm Morse, Barnes-Brown & Pendleton were extremely patient when a bunch of geeks tried to incorporate a company based on mathematical formulas rather than legal documents. Jon and his colleagues Donald Parker, Howard Zaharoff, and Robert Shea provided extremely helpful input on many of the matters in this book.

Our advisor, Neil Gershenfeld, gave his blessing when almost his entire team of PhD graduates gave up academia to pursue a commercial career at ThingMagic. Some of us have happily returned to academic pursuits since then.

Many of my friends and colleagues have read various drafts of my manuscript and helped me get it in shape: Thomas Weber, Beth VanPelt, Tom Grant, Jon Gworek, Mark Schwartz, Tim Rowe, Ken Lynch, Tristan Jehan, Thomas Bryner, Viktor Schoner, and Mary Farbood. I am very grateful for your time and thoughtful comments.

My agent, John Willig, patiently introduced me to the world of book publishing, which I find at least as confusing as the high-tech industry.

My editor, Casey Ebro, took on a book by a first-time author who writes in English but thinks in German. Casey's enthusiasm and commitment made this project happen. Along with Casey, I would like to thank the entire McGraw-Hill Education team, including Jane Palmieri, Cheryl Hudson, Elena Magg, and Courtney Fischer.

My publicists, David Hahn, Karissa Hearn, and Lindsey Hall at MEDIA CONNECT New York and Jeff Nordstedt at the Nimble Agency, worked enthusiastically and tirelessly to market this book to the world.

My parents, Adelinde and Konrad Schoner, generously raised me to travel far and explore everything the world has to offer and then proceeded to provide many decades of transatlantic long-distance support. I have been blessed with their unconditional love along with that of my extended family, including my siblings and their spouses, Ellen and Thomas Bryner, Viktor Schoner, and Julia Schmitt; my many nieces and nephews (whose names are hidden in the book); my parents-in-law, Murako and Mohamad Farbood; and my sister-in-law, Mina Farbood, and her husband, Kirk Simon.

My wife, Mary Farbood, has tirelessly supported me with love, understanding, and encouragement despite her own demanding career. We started dating a few months before the incorporation of ThingMagic, and our first child was born during the final editing of this book. You had to endure both the company and the book project, namely, deal with long periods of my mental or physical absence. I thank you from my heart.

CHAPTER 1

The Entrepreneurial Dream

It's kind of fun to do the impossible.
—WALT DISNEY (1901–1966)

In the summer of 2000, four fellow graduate students and I, all with a decade or so of higher education under our belts, started to panic about what to do after our then imminent graduation from MIT. We were all going to get PhDs, which meant that we had to retire from grad school. A PhD is the one terminal degree that is truly terminal: no more hiding in fancy labs free of real-world worries; no more projects that were supposed to save the world, but—honestly— were mostly just a lot of fun to work on. We sat on the lawn in front of the Media Lab and decided to start ThingMagic LLC, a technology design and prototyping firm. Unlike most successful entrepreneurs who leave school to found a company, we founded a company because we had to leave school.

We decided to work out of a garage, just as all the other successful tech entrepreneurs in the twentieth century had done. It seemed like the right thing to do, and it was going to be free. We spent a few hours cleaning out the messy shed that belonged to the only homeowner

1

among us, put up shelves and benches, pulled a network cable from the main house, and started designing circuits the very same day.

At the time, entrepreneurial activity was imploding in the dotcom bust (Figure 1.1). Investors and investment dollars had all but disappeared. With funding nowhere to be found, we started working on development projects for large corporations in exchange for cash.

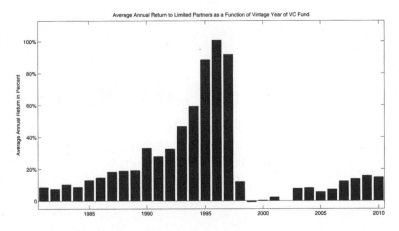

Figure 1.1: Average annual return of venture funds (net of fees, expenses, and carried interest) as a function of the vintage year (inception) of the fund. Data source: Cambridge Associates LLC in collaboration with the National Venture Capital Association (NVCA), *Venture Capital Returns, 1981-2013*, Arlington, VA, http://www.nvca.org.

To stay alive, we saved money in every aspect of our business. Our salaries were way below market, and the founders and management team had to live with a substantial amount of deferred compensation. We paid just enough bonuses to avoid the turnover of our most valuable employees. My cofounders never forgave me for purchasing what they considered test equipment unworthy of their talents: *used* oscilloscopes and computers with the Enron property sticker still on them, exchanged for a few dollars at a sketchy highway rest area. We did not travel unless someone else picked up the tab. I went as far as sharing a hotel room with a customer. Never do I want to get that close to a client again.

In managing the company through this period of bootstrapping, I earned the reputation for being an intolerable miser. I was Scrooge.

In my defense, we did operate profitably during our bootstrapping days, an accomplishment we could not repeat for a long time following the infusion of venture capital (VC). I remember those early years as a time of extreme improvisation and near-death experiences. Anyone who knew about our quarterly, or should I say weekly, numbers felt pressured by the uncertain cash flow. Yet, this early period was one of the most productive in the life cycle of the company. Frankly, I have never worked as hard before or since.

While bootstrapping the venture as a services company, we slowly built a product portfolio in *radio frequency identification* (RFID), an area in which we happened to come across a few funded projects. We didn't choose to focus on RFID technology; it chose us. Then, in 2004, Walmart and other major retailers in the United States and Europe issued their *RFID Mandates*. In essence, the retailers ordered their suppliers to apply RFID tags to all shipments. Walmart would no longer accept products from vendors that did not comply with the directive.

The technology industry stood in awe. What an unprecedented and unbelievable opportunity! The largest company in the world was mandating the use of a specific technology that hadn't even been fully validated. Even by conservative estimates, the RFID industry was poised to ship tens of billions of tags annually, in addition to supplying massive numbers of reader infrastructure, software systems, and business process reengineering projects.

For ThingMagic, the rapid market development was both a blessing and a curse. We had already established ourselves as a brand within the industry, and our name recognition and influence far exceeded our small size and revenue. We had bet on the right horse without too much thought, and we found ourselves in the best imaginable position to reap big profits and personal riches. On the flip side, the prominent opportunity quickly attracted competitors and capital. New startups popped up by the dozen, and within a few years, more than a billion dollars of investment had been poured into our industry.

Later that year, one of our startup competitors was acquired for a proud sum north of $200 million. We should have followed our competitor's example and tried to sell, but instead, we used the positive investment climate for RFID to abandon our bootstrapping ways and raise capital.

Up until then, we had pursued a licensing strategy for our RFID reader designs. A couple of large multinationals had paid us in exchange for the rights to make and distribute our technology. That strategy was working out well initially, but it became constraining when the licensees insisted on exclusivity, didn't aggressively market the devices, or couldn't respond to our own needs for manufactured products. We felt compelled to start our own manufacturing effort, which didn't exactly sit well with our licensee partners.

Should we have been surprised that manufacturing products is an expensive and risky undertaking? If you build inventory and the customers don't buy it (despite their earlier assurances), it is a problem. If you don't build enough and they want to buy, it is also a problem. Manufacturing the right number of products sounds like a simple forecasting exercise, but boy, is it difficult to guess right! We learned the hard way that bootstrapping without significant cash in the bank and manufacturing are not exactly compatible strategies.

In 2005, we bit the bullet. We changed our legal structure to ThingMagic, Inc., accepted more than $20 million in VC money, and hired a full-time CEO. The hype in our industry allowed us to raise money at a very attractive valuation, one we would never match again in future funding events. Overnight, we found ourselves in the middle of what, in hindsight, was our very own RFID bubble.

Following the capital infusion at ThingMagic, I was dumbfounded by how much our little company needed all of a sudden: annual external audits; extended liability insurance; market rate salaries; bonuses; trade-show booths ($100,000!); PR representation; volumes of legal documents; and key-person insurance. We were a *real* company all of a sudden, and being *real* turned out to be *really* expensive.

Only a few days after our Series A funding round closed, I heard one of our VPs state in a meeting: "We used to do things the cheap way. Now we are doing them the right way." I was quite offended, given my role in establishing a culture of frugality. The comment was unfortunately paradigmatic for the behavioral change in the company: every group and department felt an immediate need to spend, and hence, the money raised disappeared much more quickly than any of us had anticipated. A couple of years later, saving money was once again a top priority, and the VP quoted above had long ago left "to pursue other opportunities." While the meaning of "doing things right" changed multiple times in the life of ThingMagic, "doing things cheap" was usually spot on.

Venture capitalists and boards of directors like to use established formulas to get their arms around the unpredictability and complexity of startups. They are particularly quick to hire a team of seasoned executives to run a young company founded by inexperienced technologists. Any sign of difficulty presents an opportunity to bring in the guys *who have done it before*. In quick succession, we hired VPs of engineering, business development, sales, and manufacturing, and—just to be sure—we recruited an army of middle managers as well.

Unfortunately, it was less than two years before most of these executives had to leave. They were all very good professionals, but they could not deal with ThingMagic's specific challenges. Their prior professional experiences had not prepared them for a situation that required them to lay off people (rather than bleed to the tune of millions a quarter), change the business model dramatically (rather than design boxes that nobody buys), visit customers personally (rather than sending underlings to manage zero-revenue accounts), or shift production to China (rather than manufacture at the shop around the corner).

The boom in the RFID market was followed by the RFID bust. The industry that was going to take over the world, in which we were so well positioned, belly flopped in a heartbeat. The potential customers that had pushed us into developing this new technology all of a sudden were in no rush to spend actual money on our marvelous inventions. We had just finished staffing up to more than 60 people when we were forced to lay off almost half of our dear colleagues—without doubt the most painful period in our history.

I used to tell our disgruntled, overworked, and underpaid employees that we would be doing a bad job if we had every single resource we needed fulfilled and if we were not operating on the edge of the impossible. A healthy small business needs to manage with fewer staff than a conventional corporate approach would suggest. Surprisingly, this insight is not obvious to many startup employees, and the few who do get it easily forget it in the midst of their daily routine and workplace anxieties. Ironically and sadly, a round of layoffs silences even the biggest complainers. The survivors understand the principle that the company goes through a reduction in force in order to continue operations, preserve shareholder value, and protect the remaining jobs.

The downturn in our industry was accelerated by certain patent holders who used the opportunity of intense market interest to foment fear of an intellectual property (IP) war in RFID and to extort exorbitant and unjustified licensing fees from vendors. Ultimately, these patent holders never made much money, but their actions helped kill the interest in the very RFID products they were trying to collect royalties on. Faced with the prospect of an IP battle, many customers of the technology simply said, "Thanks, but no thanks." Since those early days, a second wave of patent trolls emerged who clearly did not learn the simple lesson from the earlier situation: when nobody is able to sell anything, nobody makes any money!

Unlike many of our competitors, we were able to adjust our business model and survive the RFID bust. We reduced our expenses and staff, and we focused on the development of technically advanced original equipment manufactured (OEM) components, which our larger competitors weren't interested in. Nevertheless, the A-round capital was soon depleted, and we found ourselves trapped in a fundraising spiral, constantly looking for more cash to keep the lights on.

Over the following three years we negotiated a number of convertible bridge loans from our investor group along with multiple new equity financings. As our de facto valuation was going down, the financial instruments became more and more creative. Our investors were patient and supportive. The RFID crisis had affected startups in our industry badly, but the crisis had largely gone unnoticed by the greater financial and technology communities. Investors continued to enjoy flexibility and the ability to protect individual portfolio companies, including ours.

We thought we had weathered the worst. Then in September 2008, the world economic crisis struck. It seemed as though commercial activity had come to a complete standstill. While previous crises, including the dot-com bust, had been more or less contained within specific industries, this was the mother of all crises, and it affected every market, region, and company, as well as everybody's personal livelihood.

Some of ThingMagic's VC investors ran into financial and leadership trouble themselves. In one case, there literally wasn't an investor representative left to look after us and actively manage the many millions they had put at risk. Banks started to go to extreme measures, such as foreclose on venture-funded startups, a fate that we were fortunately spared. No institution was safe, and capital came at an extra hefty premium, if there was any to be had at all.

There was one saving grace in all that misery: VC investors have no tolerance when their peers run out of money or are unwilling to support a troubled portfolio company. Those investors who don't play along and coinvest in follow-up financings are quickly demoted, and their preferred stock holdings are converted to common shares. Miraculously, ThingMagic was able to reduce its outstanding preferred stock, while raising follow-up funding rounds. As a management team, we had sudden and unexpected leverage coordinating the fight between those investors who remained stalwart in their support and those who had lost interest.

The world economic crisis resulted in modest demand for IT and RFID equipment, the benefits of which we started to feel in the years following the meltdown. In 2009, the founders and board decided that it was time to sell the company. Our investors were reluctant to pour any more money into ThingMagic, and the founder group had been whittled down to a tired and disillusioned bunch of three, some of whom barely talked to each other anymore. There was also little indication that the RFID market would recover from its curse anytime soon.

We hired an investment banker to bring a quick resolution to the matter, only to find ourselves struggling through another dramatic 18 months before we could conclude a transaction. During that final period, we lost yet another cofounder (at the worst possible moment, no less), we had a falling out with our lead investor, and we learned that an M&A negotiation doesn't bring out the best in people. The screaming matches among otherwise perfectly reasonable business-people would have been comical, had there not been so much at stake.

In the fall of 2010, we finally closed on the sale of our little company, more than 10 years after we first incorporated. We sold the venture that we and the market had transformed many times over to Trimble Navigation, a billion-dollar public technology company headquartered in California. In the end, our investors were reasonably content and glad to move on (Figure 1.2). We had negotiated hard to protect our staff and the integrity of ThingMagic beyond the acquisition. All of our employees were given secure jobs, to the extent that there can be job security in the technology sector in our day and age. Personally, I did not make enough money to buy myself the proverbial private island, and I expect to be working for a living for years to come.

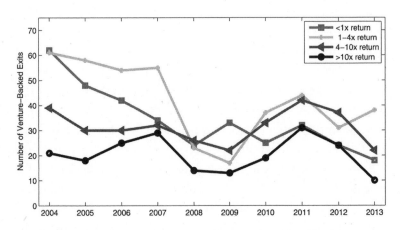

Figure 1.2: U.S. venture-backed exits and returns on exits, 2004–2013.
Data source: National Venture Capital Association (NVCA), *Venture Capital Statistics, 2000–2014*, Quarterly Reports, 2004–2013, Reports and News Releases, Arlington, VA, http://www .nvca.org.

Three of my cofounders had left ThingMagic by the time it was acquired, and even I had considered jumping ship many times during the life cycle of the company. In the end, the founders and key managers who did stick around were instrumental in the outcome of the acquisition and its impact on our employees. Naturally, talk of money dominates M&A negotiations, and yet there is a human component to this transition that requires champions lest it be forgotten. Startups that don't sell at multiples of their invested money have no contractual safety net built in for their employees. A founder who is no longer there cannot watch out for the interests of the company and the staff or her own legacy.

After the acquisition, I was put in charge of integrating ThingMagic's technology with the parent company's product offering, essentially operating as an in-house corporate entrepreneur. In many ways, the post-acquisition months and years resembled the early days of Thing-Magic. We were asked to apply our technology and make it commercially successful in a highly resource-constrained environment. To our surprise, being part of a large public corporation is hard work as well.

In summary, neither world economics nor the RFID industry nor ThingMagic's internal dynamics have been kind to me and my fellow ThingMagicians. Yet we came out the other end of the tunnel and created products and a brand that have a good chance of living on for many years.

Why have I written this book? I spent a decade of my life trying to make a technology company successful in the face of every imaginable obstacle and catastrophe. I'm sharing that experience in the hope of helping high-tech entrepreneurs get through their own difficulties with the fewest scars possible. I have also reached midlife, and I figure that authorship is a more constructive outlet for extra energy than buying a Porsche or getting a mistress.

This book introduces entrepreneurs to the process of starting, financing, and selling high-tech ventures with a particular focus on typical crisis situations at every stage of the natural startup life cycle.

Part I guides the reader through the initial startup process. Setting up shop is not rocket science, but it had better be done right in order to fully realize the business potential. Bootstrapping techniques are necessary early on, and they can come in handy at later stages of a technology venture.

Part II shows a path through the treacherous jungle of venture capital financings. Investors can bring success to a venture, but in many cases, outside funding can be fatal to the interests of founders and employees. If capital is needed to build the business, founders and managers should be prudent about how much to take, from whom, and when.

Part III offers advice on how to sell a technology company profitably despite underwhelming financial performance, disgruntled employees, and bad market conditions. Keeping the team intact and looking strong are two of the most important tools in lining up a successful exit.

Many technology entrepreneurs end up watching their companies die, wondering what went wrong. Many venture capitalists see their portfolio companies burn through millions without generating a single dollar in return. Many employees of small tech firms find themselves looking for a new job every other year without getting anywhere close to a financial windfall. It doesn't have to be this bad! With the methodology presented in this book and reasonable expectations on the part of the various stakeholders, many a startup could be spared an unfortunate end.

In this book, I address the technology entrepreneurs who are leading their companies to success, somewhere between gigantic flops and spectacular initial public offerings (IPOs). I'm hoping to help tech company founders be successful, even if they operate in a difficult market, deal with economic crises, fall out with their cofounders, or end up in bed with hostile investors. Startup books like to introduce technology entrepreneurship as a neat chess game, in which the right strategy and skill lead to inevitable success and riches. In reality, running a tech company is more like guerrilla warfare: the best planning is made obsolete by events totally outside your control. Strategy is important, but so are opportunistic tactics to keep the company afloat when earlier assumptions don't pan out.

PART I

BOOTSTRAPPING:
VENTURE CREATION
OUT OF THIN AIR

CHAPTER 2

Should I or Should I Not Venture into Entrepreneurship?

We know what we are, but know not what we may be.
—William Shakespeare (1564–1616)
Hamlet

Tech entrepreneurs enjoy a career second to none in excitement, opportunity, and reward. With the exception, that is, of those founders among us who waste our best and most formative years implementing doomed business ideas, who lose their relatives' life savings in ill-conceived ventures, or who can't deal with the tremendous emotional ups and downs of startup life.

Before you give up on less volatile career options, consider these all-important questions: Why entrepreneurship? Why now? And under what conditions? If it were only a matter of money and financial upside, the answers could be easily calculated. Yet the financial reward is only one of many issues to be considered, and—without getting ahead of ourselves—it is by far not the most important one. Indeed, if you are primarily motivated by monetary gain, you are setting yourself up for serious disappointment in the event that your company does not hit a billion-dollar valuation.

TIMING

Contrary to urban myth, only a small percentage of technology entrepreneurs are very young when they start their first company.[1] Only 15 percent of all high-tech entrepreneurs found their company before the age of 30, and only slightly more than half of all tech founders are under 40 when they start their business. The average tech entrepreneur lets 15 years elapse between receiving her terminal degree and founding her first company.

Most surprisingly, only 0.9 percent of all U.S. technology companies are founded by entrepreneurs without a bachelor's degree before the age of 25. Given this statistic, the success of famous college-dropout entrepreneurs, including Bill Gates, Steve Jobs, and Mark Zuckerberg, is even more remarkable.[2]

Numerous successful entrepreneurs have started companies at the beginning of their careers, toward the end of their professional lives, and anywhere in between. For many of us, circumstances trump careful strategic thought about the right moment. For those of you who have the luxury to choose, let's have a look at the pros and cons of founding technology companies at different times in your career.

Why should a twentysomething young kid with no professional experience to her name consider entrepreneurship? Why could it possibly be a good idea to start a company right out of college or grad school?

- **The blessings of poverty.** First of all, if your post-school entrepreneurial endeavors stumble, you won't fall very far. Your financial needs are probably as low as they will ever be. You share an apartment in borderline sanitary conditions, and your family is very small—as in one-person small. You do not carry the financial baggage that will burden later years of your life. In addition to your personal financial flexibility, you don't have the expectations of an established professional.

There will be time to make money, but for now *being poor* is just fine. You secretly hope you will be following in the footsteps of Steve Jobs, who reflected: "I never worried about money. . . . I went from fairly poor, which was wonderful because I didn't have to worry about money, to being incredibly rich, when I also didn't have to worry about money."[3]

- **Smart as you will ever be.** Right after leaving college or grad school, you have the advantage of being up-to-date on the latest and greatest in science, business, or technology. The in-depth knowledge you acquire during your secondary education is truly special, and this *freshness* will never return to the same degree later in your career.

  It turns out clever employers systematically exploit this phenomenon by aggressively recruiting the brightest of the brightest right out of school. As an entrepreneur, you have the opportunity to get the most out of your freshly gained (and paid for) education for yourself, which is a more rewarding prospect than letting some consulting or financial services outfit suck your brain dry 24/7.

- **Academic farewell gifts.** Coming out of school, you may have firsthand access to an idea, a technology, or a patent that lends itself to commercialization. Professors or scientists at times come across a new technology with commercial potential and not too many strings attached later in their careers. However, most professionals encounter this opportunity only once, and that one time is in grad school or right after completing it.

- **The innocent enthusiasts.** Finally, upon completing your education, you are more likely connected with peers who enjoy a similar free-spirited attitude, personal situation, and energy. You can find your cofounders among like-minded graduates who will go about the great adventure of founding a company with an innocent enthusiasm that professional experience will almost certainly destroy.

As a young university graduate without real-life experience, you have to rely on good common sense and your education when it comes to making business decisions. If you are lucky, you will outwit your own inexperience while you are learning on the job. Not very comforting? Here are some good reasons why you might want to become a company founder later in your career:

- **Nobody to learn from.** The downside to being an early-career founder is that you will spend the most formative years of your professional career in a situation without direct mentors and role models. As an entrepreneur, you are your own boss, and nobody can tell you what to do. As great as that is, it means that you will have to learn from the people you are doing business with or from those who are reporting to you. Be prepared for your business partners to exploit your lack of experience and for your underlings to look to you for guidance rather than advise you on what to do.

- **The benefits of gray hair.** Years of professional activity have established you in a field and have let you build a reputation. This should allow you to hit the ground running when you finally become an entrepreneur. Furthermore, you'll have had practical experience in real markets, with real customers, and with real technologies. When you spot an opportunity to create a startup, you can vet it against everything you know about your field.

- **Good fallback options.** If you fail in an early-career startup endeavor, there isn't really a career that you can go back to. You may just have to start all over again. Later in your career, on the other hand, you'll have two fallback options: either you retire or you pick up your former career where you left off.

- **Taking a career risk with a big bank account is so much more fun.** Finally, the flexibility you enjoyed right out of college may just come back to you later in your career. You will reach a point where your kids have left the house, you may be able to survive for a while without regular income, and you can once again commit more than 40 hours a week to your business.

As much as we would like to schedule our entrepreneurial endeavors as part of a 50-year career plan, life usually gets in the way. If your decision to found a company emerges from chaos in your personal life or certain peculiar circumstances, don't resist the urge. Take advantage of opportunistic situations, as counterintuitive as doing so may be: if you were just laid off, take a holy oath never to work for someone else again, and set up shop; if you invented a great technology in grad school, take it through commercialization, no matter how much your parents object; if you are having a midlife crisis and need a dramatic change, rest assured that founding a company—successful or not—will give you just that.

ThingMagic was founded by five guys who didn't know what else to do, who had no particular technology or product in mind, and who turned themselves into entrepreneurs during the worst economic climate for technology startups. Talk about bad timing!

CAREER PLANNING

Tech entrepreneurs come from all kinds of educational backgrounds and fields of study (Figures 2.1 and 2.2). Since your education doesn't really define your professional identity as an entrepreneur, the big question is, what does? How are you going to introduce yourself to a potential employer after your first startup, successful or not?

Specialist or Generalist?

Unlike a typical entry-level corporate job, entrepreneurship offers you two options. Either you develop yourself into a *specialist* by sticking with what you have learned in school and adding experience within the particular field, or you turn yourself into a *generalist* with experience in many different areas.

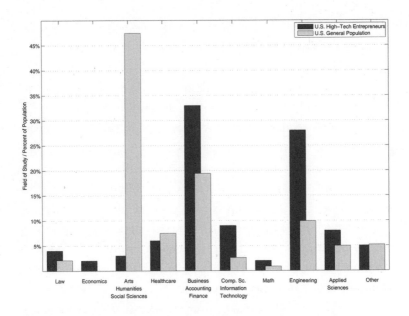

Figure 2.1: U.S. tech entrepreneurs by field of study compared to the general population. Sources: Vivek Wadhwa, Richard Freeman, and Ben Rissing, *Education and Tech Entrepreneurship*, Kauffman Foundation Technical Report, Kansas City, MO, 2008, http://www.kauffman.org, and the 2002 Census, U.S. Bureau of the Census, Washington, DC.

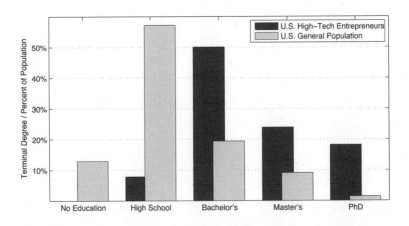

Figure 2.2: U.S. tech entrepreneurs by terminal degree compared to the general population. Sources: Vivek Wadhwa, Richard Freeman, and Ben Rissing, *Education and Tech Entrepreneurship*, Kauffman Foundation Technical Report, Kansas City, MO, 2008, http://www.kauffman.org, and the 2002 Census, U.S. Bureau of the Census, Washington, DC.

The Dilettante, Know-It-All Fallacy. Personally, I ended up in the generalist camp. Over the years I found myself doing (in this order) general management, acquisition and business development, project management, technical management, manufacturing, product marketing and sales, and then back to business development. In an early-stage firm, it's all hands on deck, and you have an opportunity to help out with just about every function of the business.

As exciting as a prolific and rotating job description can be, it is easy to lose one's way and end up as an amateur and dilettante in all things. Startups tend to lack role models and experts for many of their corporate functions simply because experts are too specialized and expensive. Consequently, the young founder has nobody from whom to learn. Wearing a gazillion hats, she also does not have the time to systematically learn, so her expertise remains superficial in all matters.

Realizing that having the ability to do many different jobs may not be enough for a sustained career can be painful. When we were preparing ThingMagic for sale, one of the serious bidders had every interest in hiring my two remaining cofounders but no interest whatsoever in keeping me. As the acquiring CEO put it: "We are interested in the technology and in the engineering passion of the *technical* founders. We could care less about the few millions in revenue Bernd has acquired." The situation was a wake-up call for me. Not having done serious engineering work for years had closed certain doors that would be hard to reopen.*

If you consider yourself a *generalist*, try to cultivate an area of special expertise. It will help you greatly with your post-startup career prospects. When it's all over, you want to be in a position to tell your future employer that you gained an abundance of experience through your startup activities, including all the vital functions of a young technology company. However, at the same time, you are most experienced, most skilled, and—hopefully—most interested in one particular area and one particular job description. It doesn't matter whether your specialty expertise coincides with your educational

---

* In the end, the deal fell through.

background. It only matters that you are passionate about it and that you are good at doing it.

The Mad-Scientist, I'm-Above-Such-Banalities Fallacy. Fulfilling many different roles, serially or in parallel, is not everybody's thing. Many a founder consciously or unconsciously chooses to stay close to her *trade*, be it technology, business, or some other function. As a *specialist*, the founder in a tech startup develops her bread-and-butter skill set. Finding a job afterward will be relatively easy.

On the flip side, the specialist founder isn't expanding her horizons in a way that will allow her to use the startup experience as a springboard for a dramatic career change. There is nothing wrong with being a bench engineer, but don't expect to be named CEO at the next company you work for.

As a *specialist* you best maximize your startup experience and *marketability* by exposing yourself to as many activities outside your own field as possible. Take a proactive part in the decision-making process, join the board, develop a public profile through speaking engagements and publications, or become the trusted go-to person of the CEO in all matters. By all means do not turn yourself into the geeky hermit who can't be bothered with anything but technology and science.

When it's time to look for a new post-startup professional opportunity, you want to be in a position to state unequivocally what your specialty expertise is. At the same time, you should make a credible case that you know how startups *tick* and you are used to working in the context of an operation with a commercial purpose.

Professional Foundation Work

In most professions, the first decade after the individual's terminal degree lays the foundation for a happy middle-class life, a productive career, and retirement. Lawyers invest a good many years to become partners, academics earn tenure, and investment bankers make enough money to retire right away.

As an entrepreneur, you have to create your own rules and metrics of success for those years. You need to embrace this task diligently, rather than dream that your startup will make you independently wealthy before the age of 30. If you count on the latter to happen, you are setting yourself up for financial ruin or a severe midlife crisis or both.

Entrepreneurship is an option for individuals of any gender, any racial background, any nationality, any industry, and any educational background.[4] Pick your field first, and then decide whether to pursue a *normal career* or set up shop yourself. If you choose a profession you don't like, running your own company is not going to make up for the simple fact that you don't enjoy what you are doing. If, on the other hand, you are passionate about your field, that enthusiasm will make being your own boss so much more enjoyable (Figure 2.3).

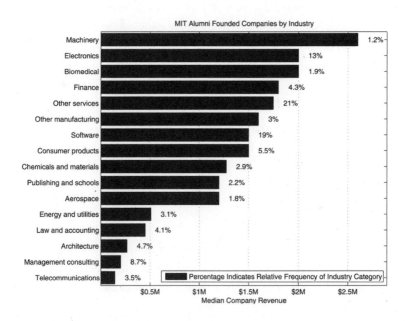

Figure 2.3: Companies founded by MIT alumni by industry sorted by median revenue; labels indicate the percentage of companies in each category. From Edward B. Roberts and Charles E. Eesley, "Entrepreneurial Impact: The Role of MIT—An Updated Report," *Foundations and Trends in Entrepreneurship*, vol. 7, no. 1–2 (2011): p. 35, Table 2.8. Published with permission of Edward B. Roberts.

Your first startup may take anywhere from a few months to a few decades to mature. If the company happens to exit on the short end of that range, what you did personally is largely irrelevant for your next professional endeavor. If, on the other hand, you end up spending multiple years on this first job, how you spent the time and what kind of name you made for yourself do matter greatly for future prospects. Reflect on this early on and frequently. Time passes faster than you may think, and what looks like a quick in-and-out adventure can turn into a long-term engagement before you know it.

MONEY, UNCERTAINTY, AND WHAT REALLY MATTERS

Entrepreneurs are commonly seen as the quintessential financial opportunists. The railroad barons of the nineteenth century, the oil tycoons of the early twentieth century, and the technology entrepreneurs of the last 30 years all amassed gigantic fortunes. Vanderbilt, Rockefeller, and Gates are all considered the richest men of their generations. Today, 55 of the 100 richest people in America are entrepreneurs.[5]

Unfortunately, these examples of financial success constitute a minuscule percentage of all entrepreneurs. In reality, only a small percentage of startups generate a big return for their founders. For the most part, the general public doesn't want to read or hear about small successes. Nor is anyone interested in the depressing stories of failed startups, unless the failures are utterly grotesque or they are noteworthy for the sheer amount of money lost.

So, while the media is reporting on the big entrepreneurship successes, 15 percent of all companies founded in the United States close their doors within the first year of operation, and 40 percent go out of business within the first five years. From the founder's perspective, the financial trade-off between an entrepreneurial career versus a more conservative path is highly debatable.[6]

Life Is Risky, Is It Not?

As much as the financial upside of technology entrepreneurship is greatly exaggerated, so is the risk associated with an entrepreneurial high-tech career. Experiencing the death of one's company is just about as awful as losing a loved one. Yet, while the emotional pain for the founder of a failing technology startup can be significant, the financial hardship likely is not. High-tech entrepreneurs are surprisingly well protected on the down side.

To start with, salaries and benefits packages in today's high-tech ventures tend to be close to market rate simply because that is what it takes to attract good talent. If a startup fails, the founder loses her job as everybody else does. However, getting laid off and having to find another job is just about routine business in the technology industry, no matter how large the employer. Unless the founder mortgaged her house to finance her failing startup (which you should never ever, ever do), she is no worse off than any employee in a large corporation who becomes the victim of a force reduction.

Why do startups look like a risky career choice then? Mathematically speaking, the individual in a startup endeavor doesn't benefit from the law of large numbers. The founder in a high-tech venture enjoys a high expected value of payout. On average, startup founders do make good money. Unfortunately, the standard deviation on the financial upside is also very high: most startup founders don't make a lot of money, but some do exceptionally well.

Venture capital investors are able to hedge their bets by spreading their investments across many companies. When most portfolio companies fail, the chosen few that are successful carry the fund. Entrepreneurs, on the other hand, are expected to put all their eggs in a single basket at any one time. They overexpose themselves to stock in their company, and they are unable to diversify. Hence they violate the most important rule of portfolio and investment theory.

Looking at the inner workings and day-to-day business of a startup, we find this fundamental and unfortunate issue at work in many places. In large companies, most situations and actions repeat themselves often even in a short period. IBM is pursuing thousands of deals at any one time. Some of them come through; many don't. Yet, at the end of the financial period, the P&L shows a nice and smooth pattern, the average of all the deals that were won. By contrast, a startup needs to focus on a small number of projects. When they don't materialize, the livelihood of the entire company is immediately at risk. This conundrum rears its ugly head in many aspects of startup life: hiring people or executives, developing product, and choosing a strategy. In a startup, you have one or two shots to get things right. If you don't, the company dies.

Venture-funded companies *buy* themselves additional breathing room. They accept investor money to fund a commercial plan or pay for past and future missteps. The money lasts anywhere from a couple of months to a few years. Unfortunately, no matter how big the round, the startup will run out of money again sooner or later.

With or without funding, what looks like a financial success story one day may turn into an utter economic failure the next. While the reverse is also possible, it is, unfortunately, much less common. Whether you call it *risk* or *uncertainty*, as an entrepreneur you are facing an uncertain and highly volatile future. You have to be able to deal with uncertainty, temporary failure, and discouragement. While your future looks very bright in general, the specifics are utterly unknown, and success may take a long time to materialize.

Indulging the Entrepreneurial Glamour

Long before a financial windfall happens, or in the absence of it, there are many reasons to be proud of the company you create. Most of them have nothing to do with your personal financial payout at exit time. To protect your sanity as an individual and your productivity as a professional, take the time early on to reflect on the things you care about and on what would constitute *entrepreneurial success* for yourself. Creating lasting employment for your employees? Commercializing a technology? Helping to shape a new industry? Inventing a new business model? Great if it works; enlightening if it doesn't?

As a company founder, you have an opportunity early in your career to impact the world in a much greater fashion and in many more ways than you ever would as an employee. You have to train yourself to recognize those moments of impact as well as the sum of all the good things you have done or will be doing as an entrepreneur. Indulge yourself in those accomplishments as long as they last. The memories rather than the riches will be the more important, and possibly the only, take-away from your endeavor. If indeed your startup does not generate a significant return for yourself, don't despair. You almost certainly created something of value. You just have to make an effort to recognize and celebrate that value.

My own industry, RFID, is filled with companies that didn't live up to what we all thought their potential was. Most RFID startups that were founded in the last dozen or so years either went out of business, were sold for pennies on the dollar, or lingered on, waiting for better times. Despite the financial misery and bloodshed, we recognize the founders of those ventures for the contributions they made.* They are the respected heroes of our industry, and yet, as far as I can tell, they all still have to work every day to earn a living. They have made their mark, whether or not riches will eventually accrue to them.

WHEN IS ENOUGH, ENOUGH?

Entrepreneurs don't like to give up. Their stubbornness makes young enterprises prevail in stormy weather and stay the course toward institutional success. While admirable, many of those ships will sink despite the most heroic efforts of the captain and her team. The question for the captain is: How much is enough? When is it time to surrender and save everybody's livelihood?

Most of us consciously or unconsciously maintain a list of minimal requirements that make up for a dignified lifestyle. The list may

---

* Kevin Ashton and Sanjay Sarma will forever be recognized as the founders of the Auto-ID Center. They had the vision and persuasion to convince more than a hundred companies to bring RFID technology to the supply chain. Chris Diorio, cofounder of Impinj, has been known as the technical soul of the standards setting process in the industry for more than a decade. John Smith, who cofounded Alien Technology some 20 years ago, gets the award for surviving the most CEOs and for burning more venture capital than any of us ($300 million and counting). Last but not least, Mark Roberti, founder of the *RFID Journal*, has been faithfully reporting on all the big, small, happy, and sad news in our industry. Mark must have an obituary on the RFID industry waiting in his drawer, but for now, the patient is still breathing.

be unrealistic and unachievable for some. The rest of us have learned that we can't have everything and that there are significant trade-offs in what to expect from life.

Create your list before embarking on an uncertain career path such as founding a startup. At any time during the startup endeavor, you should be able to confirm that the minimum lifestyle is preserved and that the startup has not taken over your life. If items on the list are being sacrificed to the company, it is time to start thinking about a career change.

Most important, when the entrepreneur's family life is suffering from the realities of a startup endeavor, the happiness of the family should take clear priority. Many a family has broken apart because one spouse prioritized the future of the startup over family life and needs. Those incidences are most unfortunate, not only for the family but also for the small company. A startup has a very small chance of survival once a key contributor goes through significant turmoil at home. Hence, the entrepreneur likely ends up in the worst-case situation: no company and no family!

The years spent in pursuit of a startup will not come back, whether or not you achieve your goals. When you look back on your twenties, thirties, or any decade of your life, you want to be able to say that you maintained a happy lifestyle that afforded you and your family the minimal amenities of your personal can't-do-without list. The riskier your goals, the more disciplined you need to be about calling it quits when things don't work out. The entrepreneur, by definition, never experiences the regret of not having tried. As a second priority, she needs to make sure she doesn't try for too long.

Entrepreneurs are easily trapped in the just-one-more-year-and-things-will-turn-around mindset. If things haven't turned around before, they may turn around in a year, but it is more likely they will not. You and your family will then be one year older and probably one year more alienated from each other.

On the flip side, there is almost always a path to take an entrepreneurial endeavor forward. If the company filed for bankruptcy, you can take it out of there. That's what bankruptcy law is for. If the venture is entirely owned by an investor group, there is usually a way to get some ownership back, simply because the investors want nothing more than to preserve the chance to make a few dollars in the future. If the product or idea was truly a red herring and did not work out, you may choose to stay together as a team and restart on a different idea. That, of course, makes sense only if the stressful experience of failure hasn't ruined the spirit of the group.

Make sure your decision to continue or abandon your venture is based on your own rational analysis of what you want and need, rather than outside opinions. Also don't just give up because you see opportunities elsewhere, be they other startup ideas or other professional engagements. Don't quit because the grass looks greener on the other side of the fence. Entrepreneurs are in constant danger of judging the company next door too positively. Startups need to look bigger and stronger than they actually are. That's the way to attract employees and build confidence with business partners and investors. While you know about all the problems within your own organization, you don't easily notice them in another company. Yet, it is most likely that the other startup is fighting with lots of problems too, and they are probably worse than yours.

DON'T LOSE SIGHT OF THE BIG PICTURE: IT'S YOUR LIFE!

Lesson 1: It is possible to start a company at any time in one's professional life. However, there is a lot to be said for founding a tech startup coming out of college or grad school.

Lesson 2: Founding a startup early in one's professional life is a career-defining move. It makes a lot of sense, but it can close the door on many other career choices. Build your own career and expertise with discipline, even in the midst of startup turmoil, when everybody expects you to do whatever benefits the company.

(continues)

Lesson 3: Don't become an entrepreneur for the money. Rather, think carefully about what metrics of success are important to you, including the pleasures of building an institution, creating employment, and being associated with a venture that made a difference. Convince yourself that you are passionate about founding your own company, irrespective of the financial outcome.

Lesson 4: Define your personal list of lifestyle can't-do-withouts. When you begin compromising on the fulfillment of those needs in favor of your startup, it is time to think about a career change.

CHAPTER 3

Startup Assets

If the highest aim of a captain were to preserve his ship,
he would keep it in port forever.
—THOMAS AQUINAS (1225–1274)

In evaluating job options after grad school, founding a tech venture myself seemed a far better deal than working for someone else's startup. Better to manage your own company to success than help someone else get rich. Better to lead your own company into ruin than follow someone else doing it. Either way, you are better off as a founder!

So, if founding a high-tech company is so great, why doesn't every engineer or technically inclined MBA do it? What do you need to bring to the table on day one to get your venture off the ground and have a fighting chance of survival? What startup assets will position you best for success down the road?

Startup assets range from early customers to patents and other forms of intellectual property to inherited software, all of which are the subject of this chapter. Since the people factor is so overwhelmingly important, Chapter 4 is dedicated exclusively to the selection of the team.

IDEAS

The public indulges stories about brilliant ideas turning into great companies. It's part of the American dream. One genius epiphany, and the next moment you're rich. Yet never has the substance of an idea *alone* been the reason for the success of a company. Hence, stories about revolutionary business ideas are typically told with 20/20 hindsight.

A catchy idea, however, does go a long way toward creating excitement among cofounders, early contributors, and investors. Your idea is the entry ticket that gets the venture creation process started. As such, make sure your idea is of reasonable quality and holds up to first-order scrutiny. Vet your idea against the following simple questions. If you can pass the sanity check, you won't be thrown out of the first meeting with potential teammates, advisors, or investors.

- **Is it just cool, or is there a market?** A big market potential is usually a good thing, but not always. You will not be able to build a $100 million business if the market is only $10 million. You also won't be attracting anyone's attention for a market that is in the low-double-digit millions. However, the billion-dollar markets tend to be highly competitive, hard to enter, vetted by other entrepreneurs, and scrutinized by investors. A nice little niche market may be just the right place to get started. You can always expand from there.

- **Is there even the slightest chance that I can get to my customers?** The fact that a market for an idea exists doesn't mean it is yours for the taking. High-tech startups are good at cracking markets that don't require large marketing budgets, that are concentrated on a small number of customers, that are regionally contained, and that don't require an elaborate channel. Entrepreneurs are notorious for underestimating what it takes to socialize a product with the customer base. This is not about you or your skills. It's just a very hard thing to do.

- **How many decades will it take to get widespread adoption?** Assuming you actually have reasonable access to your customers, how long will it take your customers to adopt your offering? Come up with a realistic and honest estimate and then multiply it by 5 or 10. You are unlikely to overestimate the time it takes for your target customers to make a move, however great the technology.

- **Will it provide happiness, pain relief, or more pain?** Customers buy things either to gain pleasure and convenience or to avoid pain. Personally, I think the latter motivation is stronger and more robust than the former.* Have you ever heard of someone rejecting pain medications because they are too expensive? When in pain, people fall for anything that promises relief.

- **Why ME?** Vet your business concept against your personal expertise and that of your team. There is no point in implementing an idea that you are not qualified to pursue. Make sure you know a reason or two why you are the right person to start a particular company.

- **Is now the right time?** If your idea hits too early, it won't work. For example, if the chosen technology is not mature or it is too expensive, you'll run out of steam trying to get it to prime time. On the other hand, if you are too late, someone else will have already done it or a new technology will have made your idea obsolete.

At the MIT Media Lab, we used to joke that ideas and demos were being recycled on a 10-year schedule. A decade was just about long enough for corporate sponsors and the media to forget and to get excited again about the same good or bad idea. Unfortunately, in the

---

* Loss aversion has been well documented in the behavioral sciences as summarized in Daniel Kahneman's recent book *Thinking, Fast and Slow* (Farrar, Straus and Giroux, 2011).

commercial world, there is at most one good moment to implement an idea. You have to hit that moment spot on for your idea to have a chance.*

Many among us feel like we don't have enough ideas, while others are seemingly overflowing with imagination and creativity. Don't be intimidated. You can systematically come up with good ideas given a good attitude. On the other hand, if you lack the discipline and persistence to think through a business concept carefully and execute it well, the greatest abundance of ideas won't do you any good. Quality and refinement trump quantity as far as business ideas are concerned.

EARLY CUSTOMERS

For a new venture run by engineers, no challenge compares in difficulty to the task of securing a customer base. *If we build it, they will come*, believe the technologists, which, of course, is not the case.

Because customer acquisition is such risky business, the more paying clients you can line up for your venture before you get started, the better. If a customer writes you a check to begin the work, you just secured the most powerful endorsement of your business model. Critical stakeholders in your venture understand that. Paranoid investors, scared employees, and aggressive board members will cut you a lot of slack if you can show early, committed revenue.

Use one of the following strategies to get customers on board by the time you are serious about the business.

Engaging Disillusioned Research Sponsors

Having poured considerable funds into university research, corporate research sponsors find themselves scratching their heads trying to turn the research and IP they paid for into commercial products.

---

* To offer just one example for this timing issue: Students at the MIT Media Lab conducted experiments and research in wearable computing in the mid-1990s, earning them the designation *cyborgs* and quite a lot of media coverage. Yet, it took another decade and a half before those concepts became a more mainstream technology (notably through the Google glass project) with the promise of being the next big thing in consumer electronics.

Universities are good at doing research and taking technology to a prototype or proof-of-concept level. They are not good at all at designing actual products guided by product marketing and engineering specifications.

This is where you, the recent or soon-to-be graduate, come in. You know the technology from your work in the university lab, and you have a good idea about the sponsor's needs. You can negotiate terms with the sponsor before even leaving the university. Make sure that IP ownership of your work product is shared in some fashion between the university, your new venture, and the customer-turned-corporate sponsor.

Helping Niche Customers

Customers of large corporations at times find themselves unhappy with the level of customization they can get. The specific product, service, or feature they need may be too much of a niche requirement to attract the attention of the billion-dollar corporation. However, what is too insignificant for your large employer may be just attractive enough for your own venture.

When you quit your job to found a business around the opportunity, watch out for confidentiality and non-compete issues.

Picking Up Obsolete Products

Companies leave stranded customers behind when they go out of business, change strategies, or discontinue product lines. You should be the one benefiting from the situation by jumping in and helping the customers through your own firm. To make this happen, you can (a) license the discontinued technology, (b) offer support for the discontinued technology, or (c) offer a service or product that avoids obsolescence of a large install base of the obsolete product. In this case, as in the last, the original technology providers should welcome your efforts since you will help salvage their reputation and help their customers in a difficult situation.

Taking Customers with You

In certain industries, the relationships between the customers and technologists can be more important than the firm-to-firm relationships. For example, if your firm is providing software services and you have built up expertise in a domain or with particular customers, chances are the customers will want to continue to work with you, even if you leave your firm and work under your own name. This is very typical for lawyers and accountants. In high-tech professions such transitions are more difficult to achieve because IP rights and confidentiality and non-compete agreements can make it hard to compete with former employers.

Crowdfunding

Crowdfunding has become a popular method for funding the development of consumer technology. In addition to raising money, crowdfunding enables you to get support from real customers who want to buy what you intend to develop and who are willing to put their money where their mouth is (see the section "Crowdfunding" in Chapter 5).

PATENTS AND OFFENSIVE IP STRATEGIES*

An offensive IP strategy at its best establishes a monopoly position for the holder of a set of patents. Anyone who wants to make, sell, or use the patent-protected invention needs to work with the patent holder or get a license. In the biotech and pharmaceutical industries, these monopoly situations are indeed common. New pharmaceutical products are immediately and securely protected by patents drafted to keep the competition at bay for the lifetime of the patent.

---

* This section includes ideas on how to organize your intellectual property and patent strategy. Don't consider the content to be legal advice, but consult a patent lawyer to discuss your specific situation.

In the high-tech sector, however, the situation is more compli-cated. Products tend to include a mix of technologies and often de-pend on multiple patents. High-tech companies are used to the fact that they do not control all the patent rights to an invention or a product. This is particularly true for products that comply with an industry standard (Figure 3.1).

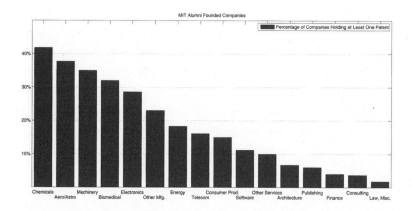

Figure 3.1: MIT alumni entrepreneurs' survey: percentage of alumni-founded firms holding one or more patents by type of firm. From Edward B. Roberts and Charles E. Eesley, "Entrepreneurial Impact: The Role of MIT—an Updated Report," *Foundations and Trends in Entrepreneurship*, vol. 7, no. 1–2 (2011): 1–149, p. 40, Figure 2.11. Published with permission of Edward B. Roberts.

Rather than forcing each other out of business, high-tech com-panies license their patents to each other in order to come up with viable offerings. The individual patent serves as an important bargain-ing chip in the explicit or implicit IP negotiation, but it rarely is used or suffices to shut down a competitor. Industry participants quietly and implicitly acknowledge that there are other patent holders in the industry but that enforcement and litigation would be in nobody's best interest.

In the RFID industry we have had multiple attempts by indi-vidual companies and patent holders to dominate the IP landscape. These hostilities have done significant damage to the market develop-ment, which has contributed to the RFID downturn following the retail supply chain hype. Monetization of RFID patents worked only when the patent holders showed restraint in setting reasonable terms and royalty rates.

Startup executives and individual inventors like to complain about the inequities of the patent system, about the difficulties in defending a small startup against IP enforcement by large and evil corporations, and about the ruinous cost of filing patents. No question, filing patents costs a ton of time and money. However, the patent system offers many degrees of freedom to implement effective patent strategies. The rules allow for effective IP programs that do not require the budget of a Fortune 500 company. As a small company technologist, you need to exploit these strategies to the fullest extent possible in order to stay competitive and to defend yourself against hostile actions from competitors.

The Magic of Patent Continuations

Utility patents are the most common patent category. They may be granted for inventions of a new method, machine, device, process, or chemical compound. Utility patents consist of two major elements: the description of the invention and the specific claims of the inventors. The description has very few formal requirements, but the claims are restricted in scope. A description can include multiple inventions, while the claims section is limited to one specific invention. Once a description is filed with the patent office, multiple sets of claims can be filed in separate utility patents based on the same description. All of these filings can claim the priority date of the first filing, as long as the chain of priority remains unbroken.

If certain conditions are met, new patents can be filed and awarded many years after the description was drafted and filed, while claiming the original priority date. In practice, the inventors can amend or change the claims in a filing and resubmit until the patent formally issues. Furthermore, inventors are allowed to file patent continuations: continuations have a new filing number and a new set of claims, but they use the identical description of the earlier filing. The new filing essentially opens up a new thread, which in turn can be the basis of a new continuation and so forth.

In short, the patent law allows for an arbitrary number of patents for little more than the price of one. While filing fees and maintenance fees are due for every new continuation, little patent attorney time is required to draft a new set of claims under the direction of a clever inventor. All of the strategies outlined in the following are in one way or another enabled by the principle of patent continuations.*

Provisional Patents: The Poor Man's Quick and Cheap IP Protection

Let's assume you just spent your last startup dollar on a key technical invention and the development of your first prototype. You are eager to show your baby to potential investors and customers, but you have neither the money nor the time to protect the technology in a utility patent filing. What to do?

A *provisional patent filing* allows you to get any material on file with the patent office without format constraints. So you assemble and file every document you have ever written about your technology, including manuals, test results, lab book pages, design documents, figures, and PowerPoint presentations. The more detail, the better, as long as the material is focused on technical aspects rather than your business plan! The goal is to include enough information to enable someone of skill in the field to practice the invention without undue experimentation. It doesn't take much time, and it doesn't take an expensive attorney to do the work. The filing fees are comparatively minor as well.

Once you file a provisional patent, you have the right to file a more carefully thought out utility patent within 12 months. The utility filing claims the earlier priority date. Any invention that was

---

* *Continuations-in-part* are variants of patent continuations. They include the original description and some additional material. Consequently, the priority date of the individual claims depends on whether those claims require the support of the earlier descriptions or the later additions.

mentioned and supported with sufficient detail in the provisional remains patentable as long as a utility patent based on the provisional is under examination at the patent office.

Kitchen-Sink Patents: Not Pretty, but Efficient!

The patent law restricts the number of claims in a single patent. However, it does not restrict the number of patents and claims filed under the same description. Any invention supported by a patent description can ultimately be claimed. As a cash-strapped startup, you can use this principle to draft a *kitchen-sink patent*: you write a comprehensive description that includes any technical invention and business activity of your startup plus anything that you might possibly do in the future. You file that monster description with a first set of claims.

As you go about your business, learn what really matters, and receive funding, you file one continuation after the other, slowly building up a library of patents covering important inventions in your field. As long as the inventions were mentioned in the initial kitchen-sink description, they all enjoy the original priority date. You should, however, keep in mind that you released the rights to any invention that you ultimately do not pursue in a claim into the public domain.

Submarine Patents: Ethically Acceptable if Used for a Good Cause?

The term *submarine patent* carries a bad connotation due to its association with patent trolls. The term was used in the past for patents that would be under examination at the patent office for years if not decades. The patent owners would only let the patent issue when they knew which infringers they would go after. The term is also used for patents that are not enforced by the owners until a third party has built a significant business based on infringing technology. Patent trolls like to come forward and extort a maximum of royalties and penalties when the infringers actually have the means to pay. This practice substantially decreased when the patent office changed the term of a granted patent from *17 years from date of issue* to *20 years from date of filing*.

Defending oneself against patent trolls is as hard for a small company as it is for a large technology corporation! No patent you own can protect you against trolls. However, you can use your own submarine patent as a defensive tool against practicing patent holders.

Letting the Patent Office Do the Work for You

Short of spending all your startup's operating budget on IP lawyers, it is extremely hard to draft claims broad enough to cover as much ground as possible, yet narrow enough to not step on prior art. Fortunately, the patent examiners at the patent office are paid to figure this out for you, as long as you give them enough description and claim material to work with. You do have a duty to disclose any prior art you know about, which is the other ingredient into the lengthy negotiation between the inventors and the patent office.

The very first office action from the patent office will tell you which claims the examiner can accept, which ones he might accept if you put up a fight, and which ones are not going to fly no matter how hard you try. It is sometimes more economical to debate the issue with the patent examiner during the review process than to do all the work up front.

Alternatively, a quick patentability search prior to filing the utility application can help you identify prior art information that you can use to shape your claims and increase your chances during the review process by the patent examiner. Prior art searches are relatively inexpensive.

The Burden of Patent Protection

As great as patent ownership is, it can turn into a burden for the small technology venture. Here is why:

- **IP enforcement can be deadly, no matter which side you are on.** Enforcement of patents against a competitor can be fatal for a small company. The cost and effort of litigation is so enormous that the lawsuit will more likely ruin the company than result in a justified win.

- **There is more than one way to peel an orange.** Patent holders like to think that they have found the only approach for solving a particular problem. Yet rarely is there not another way of implementing a product, process, or service. Don't fool yourself into thinking that your competition will not find one of these alternatives.

- **Patent fees rack up quickly.** Patents are expensive to file and to get through examination. In addition, after patents issue, the patent office requires a periodic maintenance fee to keep the patents active. Those fees may seem moderate in isolation, but they can add up to a significant bill for a small company.

- **Don't be boxed in by your own patent!** Structuring your business plan around a patent or a set of patents can be constraining. If you just spent precious equity or cash on a patent, you will certainly want to see a return on that investment by actually using it. However, the business opportunity may be somewhere else. You need to be able to go where money can be made and not feel like you have to implement the technology described by your patent.

Small technology companies are in danger of seeing the world through the narrow lens of their very own technology. They are quick to implement their own inventions rather than considering the true needs of the customer or new market developments. While it is true that ventures need to focus on their competitive advantage, which is often manifested in a particular technology, it is also true that this kind of agenda can get in the way of commercial success. Never forget that the greatest technology is not going to get you anywhere if nobody is buying it.

As you come up with your startup strategy, definitely go ahead with your plans, whether or not you have patent protection on your idea. Even if you start without patents, you will come up with new inventions as you ramp up your business and product portfolio. If you don't have the money to protect your inventions first, start the business, and file for patents later, when you know what's important.

DEFENSIVE IP STRATEGIES

Many tech entrepreneurs suffer their first startup setback when they find out that someone else has patented a piece of technology they deemed essential for their business plan. While it is usually unclear whether the discovered patent indeed reads on the startup's product, whether it is valid or whether it is truly essential, the mere existence of the third-party patent dampens the enthusiasm of the founding team.

My advice to such disillusioned tech entrepreneurs: take someone else's intellectual property seriously, but don't let the mere existence of a patent stop you from pursuing a promising business idea. If your venture is unsuccessful or even moderately successful, nobody will really care, and no patent holder will waste money to sue you when nothing is to be gained. On the other hand, if your business becomes wildly successful, you will have the means to defend yourself or pay a license fee.

That said, there are a number of ways to protect yourself from the aggression of greedy IP holders. Most important, indulge yourself in the challenge of working around an existing patent. There usually is a better mousetrap to be invented!

Standards and Patent Enforcement

Recent changes in patent law and patent enforcement push patent holders to cooperate and license to each other, rather than litigate. *Cease and desist orders* against patent infringers are rarely issued anymore. Instead, courts favor licensing and cross-licensing arrangements between industry participants, especially if any of the following applies: the intellectual property conflict is taking place in an established and well-functioning industry that benefits from contributions of many vendors; an efficient marketplace has been established for end users or consumers; or the technology is covered under a standards setting process.

Standards authorities usually include provisions regarding patents and licensing in their charter. For example, all International Organization for Standardization (ISO) standards require that participating patent holders agree to license necessary patents on a nondiscriminatory basis and for reasonable fees. The rules protect technology vendors who offer products that include the standard. This is good news for small technology companies. Licensing can be expensive, but it sure beats litigating a technology giant.

Patent Pools

Increasingly, patent-rich industries are organizing themselves in *patent pools*. Multiple patent holders contribute patents considered necessary to operate a specific standard. A licensee of the pool gets access to all the rights in the pool and is protected against lawsuits from all the participants. Economically, everybody benefits. The licensees pay a moderate fee for access to many patents from a number of patent holders. The licensors earn money on their patents with very little administrative overhead. As an added benefit, patent pools establish a fair licensing rate for patents in a particular industry, which helps resolve disputes related to patents that are not part of the pool.

Antitrust laws require patent pools to issue a license to any applicant. This works in favor of smallish companies that are not bringing much IP to the table themselves. A startup can protect itself from hostile patent holders without paying exorbitant licensing fees and without engaging in costly litigation.*

---

* There are multiple major patent pool administrators worldwide, including SISVEL (UHF-RFID, MPEG audio, LTE, and other pools), MPEGLA (MPEG-2, ATSC, etc.), and VIA Licensing (AAC, 802.11, etc.). As new patent-rich industries emerge, these administrators work with the patent holders to form pools within the industry.

University Licensing Programs

Universities like to facilitate the licensing of their intellectual property for use in the commercial sector, especially when the inventors themselves are running the business. According to most such programs, the eager but poor entrepreneurs are given the rights to the patent under cash-neutral terms. Cash neutral, of course, doesn't mean free. In exchange for a license, the university expects any or all of the following:

- **Equity.** Common stock in the low single-digits percentage points should be adequate, assuming that the university's role in the venture is rather passive. At times, a research institution may spin out technology in a commercial entity, whereby its role is closer to that of a cofounder or investor. Naturally the equity compensation should be higher.

- **Royalties.** Royalty arrangements have the benefit that they cost money only when a technology is successfully used and sold. Stay away from guaranteed minimum royalties though!

Before you license your inventions back, make sure the licensing agreement with your alma mater addresses the following issues and terms:*

- **Exclusive or nonexclusive use.** The former is not necessarily better than the latter, but it is certainly a lot more expensive.

- **Right to sublicense.** The licensor will try to prevent you from sublicensing the IP to another player. Depending on your business model, this may or may not be acceptable.

- **Right to transfer and/or assign.** In the case of a change of control, the acquired company needs to be able to assign the license to the acquirer.

---

* Definitely consult with an attorney before you sign any documents. No doubt, the licensors engaged the help of an army of legal folks too.

- **Ownership of derivative technology.** The licensed technology should be a starting point for new inventions. You need to be in control and have ownership of the derivatives you invent going forward.

- **License restrictions.** In order to work with additional licensees later, the licensor may try to restrict your use of the technology to (a) a field of use, (b) specific customers, or (c) specific regions. Don't let them box you in too much! You never know where and what you will end up selling.

Many of the people you need to impress when you start a company do not have the expertise, time, or resources to truly evaluate whether the licensing and patent rights you are holding are worth anything. The perception of a strong IP position alone can go a long way! As you devise your patent strategy, keep in mind that most outsiders do not have the same sophisticated understanding of the IP landscape in your industry. This lack of insight can work against you or in your favor, depending on how well you are spinning your story.

COPYRIGHTS, TRADE SECRETS, AND KNOW-HOW

Patents are for lawyers to fight over. Trade secrets, copyrights, and know-how, on the other hand, make technology companies tick. You can run a tech company without patents, but you cannot run a tech company without technology (Figure 3.2).

Copyright: Protecting the Words, but Not the Meaning

While patents protect ideas and abstract inventions, copyrights protect hardware designs, software source code, formulas, recipes, and documentation. The moment you create any such work product, it is protected under copyright law, whether or not you actually mark it as such or register your claim.*

---

* Even though the legal ownership of a document is not affected by its copyright marking, be diligent in marking your work product as copyrighted, so that enforcement of your rights becomes a lot easier.

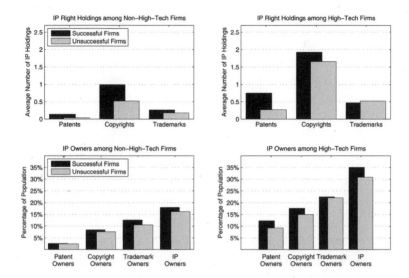

Figure 3.2: Intellectual property ownership statistics among young non-high-tech companies (left) and high-tech companies (right); categorized as *successful* (still in business or sold or exited by 2009) and *unsuccessful* (no longer in business in 2009). Data sources: Alicia Robb and E. J. Reedy, *An Overview of the Kauffman Young Firm Survey, 2004–2009*, 2011, http://www.kauffman.org and the *Kauffman Firm Survey, 2004–2009*, 2011, http://www.kauffman.org/research-and-policy/kauffman-firm-survey.aspx.

Copyright protection covers concrete technology that often is immediately applicable in a commercial business. For example, a piece of software may be immediately used in a product, while an abstract patent requires reduction to practice. When the Google founders started their commercial venture, they hit the ground running because a lot of code had been previously developed at Stanford. Naturally, Google has been improving the search engine ever since, but the foundation of its core product had been implemented and was ready to go on day one.[1]

There are a number of ways you can benefit from copyrighted work product when starting your technology venture:

- **Putting academic output to work in the real world.** If you developed a significant code base as a student or researcher in a not-for-profit lab, ask your university to let you use it commercially. Much like a license to patents, the institutional owner of the technology should be compensated with a combination of equity and royalties. If you are really lucky, the university-based research continues on in parallel with the commercial activity. Try to secure the rights to that work too, and don't forget to formalize your arrangement with the university.

- **Benefiting from another company's misfortune.** If the company you worked for developed a substantial technology base but it went out of business, the technology may be available at a low price. You might be able to pick it up in exchange for some equity in the company you are about to start with a number of former colleagues. Hopefully you and your team have learned from the earlier mistakes by the time you try again.

- **Benefiting from the fruits of your moonlighting efforts.** If you have developed technology in your free time before starting your own company, you can hit the ground running when you finally incorporate. This is the least controversial case, but there are quantitative and qualitative limits as to what you can expect from your moonlighting activity.

Before you spend money or equity on the right to use copyrighted technology, think about whether you really need the particular implementation. Code becomes obsolete and constraining rather quickly. A program that was hacked up at a university by inexperienced undergraduates may not be the best starting point for a commercial undertaking. As long as the work is protected only by copyright, you are free to re-implement the underlying concepts and functionality from memory. Just make sure you don't accidentally copy code verbatim.

Trade Secrets: Keeping Quiet Has Its Limitations

According to the Uniform Trade Secret Protection Act, *trade secret* means "information, including a formula, pattern, compilation, program, device, method, technique, or process, that (i) derives independent economic value, actual or potential, from not being generally known to . . . other persons . . . and (ii) is the subject of efforts . . . to maintain its secrecy."[2]

If you don't have the money to patent an invention, or someone else beats you to it, you call it a trade secret. Unlike patents, which are made public on a well-defined schedule, trade secrets have to stay secret, which can be a powerful tool to protecting your proprietary knowledge and competitive advantage.

Unfortunately, trade secrets provide protection only so long as they actually do stay secret. You need to be fully aware of the following realities and limitations of trade secrets in a competitive marketplace:

- **Nondisclosure agreements (NDAs) are overrated.** Employees of your company or anyone else bound by an NDA are prevented from making use of trade secrets elsewhere. However, this is mostly a theoretical constraint. Just because there is an NDA doesn't mean you can erase information from people's brains. When your engineers move on to the next job, they will certainly use what they have learned, secret or not. It's called experience.

  While courts enforce NDAs to protect unique company know-how and secrets, they generally will not prohibit engineers from utilizing their learnings and from finding employment.

- **Reverse engineering is looked down on, but it is perfectly legal.** Anyone who reverse engineers your product without breaching a contract or breaking the law is free to use what he learns in any way he wants to. For example, if your product is based on a particular circuit and set of electronic components, the information is readily available to anyone opening the device and looking at it.

- **Hiding your secrets is not easy, especially when you need to sell product.** You can try to hide algorithms and software methods in binary software images. You can cover your circuit boards with epoxy to obscure the parts you are using. Yet, at the end of the day, a determined engineer can usually figure out your secret sauce from the products you are selling.

Trade secrets are worth a lot until you are actually using them. Once they are implemented in a product, the chances that someone will be copying you are pretty high, especially if you managed to establish a thriving business.

Know-How: Nobody Can Steal What's in Your Head

Even though *know-how* is the hardest form of IP to quantify and formalize, it is the most essential asset in this category. What you and your employees have in their heads and know how to do, nobody can take away from you. Whatever the technical starting point for your venture, you need know-how to enhance, perfect, and commercialize your technical assets.

Know-how is the quintessential and necessary requirement for bootstrapping a technology company. Know-how and the people who possess it are also what investors like to invest in. If you don't have any know-how on your founding team, you will be off to a rocky start. If, on the other hand, you have capable engineers and scientists working for the company, customers and investors will be drawn to you for the expertise you offer and for the promise of great products down the road.

Draft your business plan based on what you and your team know how to do best. Every other consideration pales in comparison to the talents and technical skills of your people.

WHAT TO PACK IN YOUR BAG

Lesson 1: An idea alone is not sufficient to build a successful business, but a good idea helps focus your energy, it attracts attention, and it gets cofounders, investors, and employees excited.

Lesson 2: Value, nourish, and flatter your early clients! Nobody will be more instrumental to the success of your startup than the customers who give you money and believe in you when nobody else does.

Lesson 3: If you own patent rights to a good technology, consider yourself lucky, but don't let it get to your head. Before your patent can protect your business, you will have to build that business . . . which is hard, patent or no patent.

Lesson 4: Don't ever let someone else's patent stop you from starting a business you believe in. If you are successful, you will be in a position to pay royalties and license fees. If you are not successful, nobody will care.

CHAPTER 4

Cofounders

From this day to the ending of the world
But we in it shall be remember'd;
We few, we happy few, we band of brothers.
—WILLIAM SHAKESPEARE (1564–1616)
Henry V

The assembly of the founding team impacts the path and outcome of a new venture more than any other decision in its life cycle. By definition, the founding team cannot be changed later. The founding team, for all practical purposes, is irreplaceable. A good team of cofounders makes for a strong foundation on which difficult situations can be overcome. A bad team turns daily routine into a nightmare and will fail when the first crisis hits.

By the time we founded ThingMagic as a group of five, all of us had spent many years in the same research lab together. We had worked together on numerous projects, and we had largely been close friends. I was convinced that we knew each other extremely well and got along great. Little did I understand that the pressure, uncertainty, and existential anxiety felt in a young company can cause these very same people, including me of course, to act very differently. At MIT, we had worked in the safe environment of an academic lab and graduate student projects. At the company we had to face existential questions on a daily basis.

The stress on the team of cofounders manifests itself in a number of ways. Founders coming right out of school are figuring out for themselves if entrepreneurship is the right career move. Financially, there is no stability yet, neither for the company nor for the individual founder. Finally, startup teams don't get much reassurance from anyone—especially pre-funding—while they face a universe of challenges that could drive the most seasoned executive insane.

In light of these daunting prospects, what kind of personalities should you engage with to start your company? Which teammates should you be looking for?

ATTITUDES

For many a personality trait deemed important for an entrepreneur, the exact opposite property is required just as well. Entrepreneurs should be able and willing to listen and take advice. Yet, they also need to have the self-confidence to not change direction on every recommendation from a random advisor. Entrepreneurs need to be able to stay the course through setback and crisis. Yet, they are doomed if they stubbornly insist on the implementation of a plan that is unworkable.

Here are some of the dialectic personality traits you should develop in yourself and look for in your cofounders.

Persevering, but Realistic. Just when you think you and your venture have survived the most difficult startup phase, things really start to hurt: your personal finances are looking dire; your significant other is seriously unhappy; and friends and former colleagues have advanced in their careers and are making much more money, while working much less.

The committed entrepreneur has to endure such hardship and humiliation and does not quit until the venture is either dead or flourishes. At the same time, the entrepreneur cannot stubbornly cling to a business model that is not working out. Rather, she needs to constantly validate her concepts and approach in light of economic realities.

Concerned with Value Creation, but Financially Ambitious. The prospect of riches motivates founders to start a new venture in the first place and keeps them committed as those ventures go through difficult times. Yet, the day-to-day work of the founding team needs to be focused on the creation of lasting and scalable value. Pick your metric, be it revenue, profit, or number of users, and then work like hell to keep those metrics on target. Financial reward will follow the value creation.

Hardworking, but Focused on What Truly Matters. Much of the work in a new venture is done best by the founder herself. Nobody speaks to the customer with more authority than a founder; nobody makes the analyst feel more comfortable than a founder; and nobody conveys more confidence to the staff than a founder.

Find yourself teammates who work hard and are willing to go the extra mile, who recognize the responsibility they carry as founders, and who enjoy the glamour of the role.

Humble, but Self-Confident. Success of a new tech venture depends on the contributions, goodwill, and input from many. The entrepreneur needs cofounders and employees to complement her skill set and share in the workload. She needs customers and partners to buy her product. She needs investors who entrust her with their money. In the process, the entrepreneur will be rejected many, many times.

If humility and patience are not your strong suits, starting a high-tech venture will teach you. At the same time you need to be sufficiently self-confident as to consider disappointments statistical events, not expressions of structural problems with you or your business plan.

Adventurous, but Respectful of the Unknown. For centuries, America has been welcoming adventurous immigrants from all over the world. Today, the population growth in the United States is still fueled largely by immigration of some of the brightest and most energetic people from around the world.[1]

The people who have come to the United States over the centuries share in the willingness to start a new life and to sacrifice the security and predictability of their previous lives. The entrepreneur makes a very similar sacrifice and bet: she forgoes a safe and predictable career in favor of the potential financial upside and freedom but also the possibility of crushing failure.

Extroverted, but Self-Aware. Entrepreneurship is a social activity. Extroverted behavior helps leverage contacts, motivate employees, and convince customers to buy. If you are introverted and shy, pair up with an extroverted cofounder. If you consider yourself nerdy, find a business partner with communication skills and form the classic inventor-plus-businessperson duo. Even better, make an effort to learn how to communicate—doing so will help you find new friends, and it will magnify your technical and business skills dramatically.

Faced with the choice of a cofounder who has great skills versus one who has the right attitude, I'd always pick the latter. Attitude lasts longer, and it trumps specific skills. That said, a good mix of skills among the founding team is an unbeatable asset when things are going well and a good insurance policy for difficult times.

SKILLS AND ROLES

Skilled employees are difficult to find and expensive, whether you hire a consultant or a full-time employee. The more skill and diversity you have assembled among the founding team members, the longer you will be able to run without *buying* talent and the smaller your risk is of hiring the wrong people.

Core Roles

Try to assemble the founding team so that you can rely on a good number of core skills among yourselves. Ask yourself what you would need your buddy entrepreneur to do if your company is going through a product development cycle, a funding round, a business development effort, or a merger.

The Prima Donna Genius. Not surprisingly, technical expertise is the one skill a high-tech founding team can't do without. You need to have a genius or two on your team to get off the ground.

The genius's competency can be highly specific. She may have acquired expertise on a particular project in a research lab or university, and she possibly backs her expertise with a set of patents. In other cases the genius's broad expertise covers a particular technical marketplace. If you are lucky, your genius brings not only technical expertise to the table but also a set of commercial contacts.

Unfortunately, technology-savvy founders tend to be rather high maintenance. Chances are, you will curse your tech genius as often as you find her contributions helpful. It doesn't help that she has accused you again and again of incompetence. It bugs you considerably that, despite her bad behavior, you have to continue to be nice to the tech genius since you can't afford to lose her.

The Leader. Running a new company in a consensus-driven democratic process has its limits, especially when hard decisions need to be made that affect everybody's lives. Consensus usually requires compromise, which is not necessarily in the best interest of the new tech venture.

A founder group with a clear leader in its midst has it easier. Being the leader doesn't mean having more stock or equity, nor does it necessarily mean being the CEO. It just means that the cofounders trust one of their own and are willing to follow her, if indeed there is conflict and controversial decisions need to be made.

The Industry Veteran. Any competent marketer can study an industry, get quick insights into how it works, understand who the key players are, and identify product opportunities. However, it takes long immersion in a marketplace to call yourself an insider, to understand the subtleties of the competitive landscape, to recognize people as true assets (despite their titles), and to look through the propaganda of technical collateral and PR campaigns. That's why the industry veteran is helpful.

Watch out, though, for those veteran candidates who live in the past. Your company needs to find a new angle to solve old problems. The veteran who has the ability to define that angle with an open mind is a perfect hire. However, stay away from the we-have-always-done-it-this-way industry veteran.

The Sales Animal. Young high-tech companies are at constant risk of forgetting that they actually need to sell the wonderful technology they invented. A sales animal on the founding team helps to contain that risk.

I do not believe that a good salesperson can sell any product. However, a bad salesperson can fail at selling a good product. If there is nobody on your team who knows how to sell, hire someone who is good at it. If, however, one of the cofounders has sales animal talents, that is, she knows how to build relationships, communicate competency to customers, and hold the line when it comes to money, you've got yourself a great asset. The combination of technical insight, founder authority, and sales experience is a hard-to-beat advantage in a competitive marketplace.

The Financial Suit. Professional controllers and CFOs are readily available for hire to fill the financial gaps on your team. Remember, though, that people with financial talent often have their own agendas, and they do not come cheap. If you can put a skilled cofounder in charge of overseeing the finance function, you may enjoy some extra peace of mind, and you will save precious cash and equity.

The Superstar. In the midst of silly little tasks, such as ordering office supplies and keeping the network connection running, it is easy to forget how glamorous the role of a high-tech entrepreneur can be. The world wants to think of tech founders as superstars who are doing what the average man and woman cannot. Groom such a superstar on your team, and use her as the backbone of your marketing, recruiting, and PR strategies. Fortunately, almost any combination of eccentricity, nerdiness, charisma, and social inability qualifies a cofounder to be that star.

If the superstar and the genius are the same person, you have the perfect spokesperson on your team, while having to manage a perfectly unmanageable cofounder. Her ego will bring you either success or spectacular failure. If things get really bad, be consoled with the old adage that *any publicity is good publicity.*

Secondary Skills

When you have the core skills covered, ask yourself what you would need your cofounders to do until you have the means to hire employees or if you ever have to lay off employees. In difficult times, secondary skills will help you keep the lights on and get you through the worst:

- **Bookkeeping.** Run your books yourself, but hire someone to do the mindless data entry.

- **Project management.** Whether you have project management experience or not, you will be in charge of the day-to-day management of your early and generally understaffed projects.

- **Drafting legal documents.** Understanding and modifying legal templates can help you save tens of thousands in legal fees. Make sure, though, that the templates come from trusted and reliable sources, and know your limits!

- **System administration.** The young technical founder actually likes to take system administration into her own hands. She would not be satisfied with anyone else doing the job anyhow.

- **Technical documentation.** Writing technical documentation for the product you are about to release is a healthy and educational experience. Put yourself in your customer's shoes, and try to understand the product from the user's point of view.

- **Public relations.** By doing your own PR, you save good money, and you stay in touch with the market, the media, and the analyst communities. You can learn how to do it from example since your competitor's PR blunders and achievements are, by definition, out in the open.

- **Technical support.** You will get that first support call, and you will have to take it. I'm willing to bet that you won't have a support professional hired and in place when it happens.

- **Web programming, development, and maintenance.** These skills have become commodities. However, that doesn't mean they are free. You can save a lot of money using free, open-source tools to maintain your online presence.

- **Graphic design.** If you are skilled enough to create your marketing collateral on your own, you will have it done much more quickly than if you hire someone to lay it out. Pay a professional to design a few key templates, including a good logo, and then maintain them yourself.

And to support the technical operations of the house:

- **Programming.** A founder with experience in software architecture and software best practices will go a long way toward quality and speed in the early software development process.

- **Version and release management.** A release manager is usually not the first hiring priority, and yet a bad versioning or release process can become costly and time-consuming.

- **Circuit board design.** Hardware engineers are expensive, and that first little circuit board you need hardly justifies a full-time hire or an outsourced engineering project. It will be fun to design it yourself.

- **Inventory and SKU management.** Few companies start off with a need to manage an extensive hardware supply chain. Yet if you set up your manufacturing processes well early on, things will go a lot more smoothly later.

How Many Should It Be?

When we founded ThingMagic, we knew that five of us were a big enough team. In fact we had to turn down friends and colleagues who wanted to join us. To this day I wonder, though, if the group was actually too big and if a subset of us would have done better. Is there a general answer to the question of how many people there should be on the founding team?

Founders tend to get very focused on equity ownership before they invite another cofounder to the team. Wrongly so! While the ownership percentage you hand to a cofounder will be lost to you, the sharing of founder stock has a relatively small dilution impact in the long run. Soon your ownership percentage will drop dramatically due to issuance of preferred investor stock, CEO shares, bank warrants, stock options, and so on.

Instead of concerning yourself with dilution, consider the following objectives before you take on yet another cofounder:

- **Maximize skills.** Does the prospective cofounder add another core skill or dimension to the team? If she truly has a skill to offer, remember that hiring someone with that core skill later will be costly, yet less effective.

- **Maximize harmony, productivity, and a spirit of partnership.** Adding another person may complicate things—after all, it is another opinion. At the same time, a third team member can help mitigate tensions between two other members. Pick whichever combination of cofounders results in the most productive and harmonious environment. As in marriage, no financial consideration can trump the long-term emotional benefit of a happy and productive founder relationship.

- **Maximize equality of contribution.** If the levels of contribution among the cofounders vary, it will be harder to define and maintain a compensation structure that works long term. This applies equally to differences in skill and experience and to differences in work attitude and work ethic.

- **Find a leader.** A founding team with a clear leader will be better off than an unstructured group. A strong leader among the cofounders who makes decisive calls after getting input from everybody else helps keep the venture on track in difficult times.

IMMIGRANT FOUNDERS

As many as 30 percent of MIT's foreign-born alumni have started a company at least once in their life, versus only 20 percent of MIT's U.S.-born alumni population.[2] At the same time, the population of foreign students at MIT is steadily growing. Between academic year 1998/99 and academic year 2013/14, the number of foreign students at MIT increased by more than 60 percent, while the population of MIT students with U.S. citizenship or permanent resident status grew by only 17 percent.[3]

Most prospective foreign tech entrepreneurs leave their U.S. school with the uncomfortable immigration status of *nonpermanent alien*, including yours truly and two of my cofounders. U.S. immigration law is rather generous as far as company ownership by foreign nationals is concerned. Everybody is free to invest money in the United States and to incorporate a company regardless of nationality or U.S. immigration status. However, when the foreign national wants to work for that very entity she founded, things get complicated.

Even if you and your cofounders have permanent resident status or citizenship, chances are you will be hiring nonresident aliens. You just cannot ignore more than one-third of university graduates if you intend to put the best possible team together. Solving the immigration issue for a foreign employee is complicated and expensive. However, you can turn this unfortunate reality into a recruiting asset. Talented foreigners will choose your company precisely because other employers would not consider them or they would offer lesser packages than what they would pay a comparable U.S. employee.

As a startup, you have the degrees of freedom to customize your support for foreign job candidates. Try to accommodate the immigration needs of foreign individuals so that they come work for you. Once they have accepted a position, they will stick with the job for longer than a U.S. employee because leaving would mean restarting the immigration process all over or departing for home.

The tech industry is lobbying heavily to reform immigration and make it easier for entrepreneurs and startup employees. Hopefully, there will be some change for the better in the coming years, including the introduction of special Startup visas. Until then, you have to work with the tools and degrees of freedom U.S. immigration law is offering today.

Student Visas: You Are Welcome to Give Your Startup a Go, but You Better Hurry

The saving grace for young foreigners graduating from a U.S. university with an advanced degree is called *academic training* or *optional practical training* (OPT). Foreign students who study in the United States under a J-1 visa have the right to extend their stay in the United States after graduation by 15 months to work for a private company. There are two conditions for the extension, both of which are not hard to meet: (a) the employment has to be related to the student's field of study, and (b) the total duration of employment cannot exceed 15 months, including the time spent working under the program prior to receiving the degree.

Foreign students who study in the United States on an F-1 visa have it even easier. They have the right to extend their stay in the United States after graduation to work for a private company under the OPT program. The employment must be related to the individual's field of study and the total duration of employment under the OPT rule prior to and after receiving the degree cannot exceed 12 months. Foreign students graduating in a science, technology, engineering, or mathematics (STEM) field can apply for an extension of their OPT by up to 17 months.

In neither of these two scenarios is the startup company required to demonstrate significant business activity, which is why this work permit works so well for foreign entrepreneurs. When founding a company right after school, the new entity has nothing to show for itself. Starting from *nothingness*, the young entrepreneur builds up momentum in the company over 12 to 15 months. By the time she needs a regular work visa, the startup is up and running and hopefully in a position to sponsor the entrepreneur in much the same way as established corporations sponsor their foreign employees (even though there are significant challenges with startups sponsoring their very own stockholding founders for a visa application as we will discuss below).

Founders Living Abroad: America Wants You, but First You Have to Prove That You Are Legitimate

If you are not a U.S. citizen and you happen to be living abroad at the time you decide to launch a U.S.-based startup, setting up shop on U.S. soil is feasible but definitely cumbersome. There are two fundamental options: the E-2 visa and the L-1 visa.*

The E-2 visa allows for entrepreneurs from certain countries, the *treaty* countries,† to set up a business in the United States provided (a) they make an investment that is proportional to the size of the business, and (b) the business is majority-owned by nationals of the treaty country. For a small tech startup, this means that the entrepreneur can demonstrate an investment of $50,000–$100,000 or more.

---

* The EB-5 visa is not included here because it requires up to $1 million in investments and is subject to other constraints making it a difficult option for young entrepreneurs.

† The list of "treaty countries" or "E-countries" includes almost all countries in Western Europe and many countries in South America and Asia.

The application process is handled by the U.S. Consulate in the entrepreneur's home country. The entrepreneur has to present a business plan and participates in an interview, explaining where the funding is coming from and what she is going to do with it. E-2 visas are issued for two to five years but can be renewed indefinitely.

The L-1 Intracompany Transferee Visa has been established to allow foreign corporations to send executives, managers, and employees with specialized knowledge to temporarily work for an affiliate company, subsidiary, or branch in the United States. L-1s are easier to get than E-2s, but for an employee to qualify, she has to have worked for the foreign corporation for at least 12 consecutive months in the three years preceding the move to the United States. During those 12 months preceding the move as well as any time after her move to the United States, the employee has to work in an executive or managerial position or in a role that involves specialized knowledge about the company's products and services.

Some of the rules for L-1 visa applications have been relaxed to specifically enable a foreign corporation to create an affiliate startup in the United States. The L-1 visa for a startup is granted for one year but can be renewed for up to six additional years.

If you intend to take advantage of the L-1 route as a foreign startup founder, you have to get started in your home country first and spend at least a year to get your startup off the ground. Once the company is 12 months old and has developed into a legitimate business, you can move to the United States on an L-1 visa, provided that the operation in your home country continues on with its business. By the time you close down the business at home—should you choose to do so—you can no longer rely on an L-1 visa. Instead, you need to apply for a different work permit such as the H1-B discussed below.

One year of startup time may sound like an eternity, but time will pass quicker than you might think while you are getting all the needed structures in place. Use the peace and quiet and the lower costs in your home country to get some homework done, including patent filings, hardware prototyping, proof-of-concept programming, and product marketing. By the time you move to the United States, much of the ground work will be done, and you can immediately take advantage of the resources you have available in America, including access to venture capital.

Work Visas: Are You Running a Real Company?

Once academic or practical training is completed, founders and startup employees most commonly apply for an H1-B visa. H1-B visas are temporary work visas for foreign nationals with a degree or specialty knowledge in a specialized field.

The visa application by the company on behalf of the founder or early employee must support the arguments that (a) the foreign national will be performing a "specialty occupation" for the company and that she has an appropriate degree to do so and (b) the company is a valid commercial operation that wasn't just set up to sponsor the applicant's visa needs.

In support of the latter, the young company needs to document the validity of its business. The requirements are informal, but useful evidence includes the following: solid financial statements; a positive cash balance; revenue and profits; an established payroll process; company collateral in the form of a web presence, marketing material, and advertising activities; media coverage; U.S. employees; cofounders and management with U.S. citizenship; and investments. In short, the more demonstrable activity there is, the better.

The immigration services are not favorably inclined toward H1-B applicants who are major stockholders and sole executives of their companies. As a founder, you should find yourself a boss before you apply. Ideally, this means you hire a CEO and a number of additional executives (without immigration issues) to run the business. It could also mean you establish a strong and independent board that you report to (and that has the power to fire you should you do a bad job).

An H1-B visa is issued for a three-year period and can be renewed once. The government imposes a cap on the number of visas issued each year.* The H1-B year starts in October, but applications can be submitted starting in April. When the quota is reached, no more visas are issued until the following fall. Good timing of the visa application is therefore crucial.⁴

Immigration law requires market salaries to be paid to professionals based on their field, education, and job description. While you can starve your U.S. startup employees by making them work for equity,† you won't be able to do that with H1-B workers. Ironically, U.S. immigration law protects foreign workers from exploitation by underfinanced startups, but not U.S. citizens. At ThingMagic, I got paid a reasonable salary only when I was put on an H1-B. At the time we were nice enough to extend the salary increase to all members of the ThingMagic management team irrespective of nationality. All of us ended up receiving the minimum wage for PhDs in our field set by the U.S. Department of Labor.

Citizens from Canada and Mexico have an additional visa option. They can work in the United States on a Trade-NAFTA (TN) status provided that they are hired into an eligible job function. The TN status is good for three years, but it can be renewed indefinitely.

Green Cards: The Ultimate Legitimacy

After two three-year terms on H1-B visas, the only thing that keeps a foreign employee in the United States is a *green card application* (I-485, Application to Register Permanent Residence or Adjust Status). When a company sponsors the green card application of an employee, it has to show that it tried really hard to hire a U.S. citizen to do the job but it couldn't find anyone. The process includes the following:

- **Drafting a job description** that reflects the standard requirements of the position.

---

* In fiscal year 2014, the general cap amount for H1-B visas was 65,000. An additional 20,000 H1-B visas were available for applicants with an advanced degree (PhD or master's) from a U.S. university.

† . . . as long as you pay minimum wage!

- **Posting the position** online and in a local newspaper, despite the fact that nobody is looking for jobs in print media anymore.

- **Explaining why other applicants are not qualified** to do the job, provided that someone did actually see the posting and sent in an application.

The overall process can take two to four years to complete, or longer for candidates from certain countries including China and India.

Sponsoring and funding the green card application of an employee is costly, but it can make your company a very attractive and sticky employment option. Once the employee is in the middle of a green card process, she will tend to stay with the company and see the process through. Leaving to work for someone else would entail starting the green card process all over.

Employer-sponsored permanent residency is by far the most popular route to green cards. However, individuals have a few other options including the green card lottery run by the government and establishing permanent residency as foreign nationals with exceptional abilities. The latter option includes the *national interest waiver* and the EB-1 green card.*

I know no other group of immigrants as obsessed with immigration law and immigration status as my Indian-national friends. They worry terribly and panic over the smallest problem. Yet, at the end of the day, everyone I've known got to stay and work in America. Many of them founded companies. Immigration status is a serious constraint for a startup founder, but immigration issues do not have to be a showstopper.

---

* Another popular method is to marry a U.S. citizen. Marriage is the quickest and cheapest method, but it comes with its own issues. Since love is just about as unpredictable as the future of a young high-tech company, I would not recommend mixing the two worlds! Otherwise, you might lose your spouse, startup, and work permit all at once.

TOGETHER WE WILL PREVAIL

Lesson 1: The attitude and commitment of the cofounders go a long way toward the success of the team and the company. Attitude trumps skill, seniority, and expertise!

Lesson 2: Rounding out the skill set of the founder team is the best insurance policy for difficult times. What you can get done between the cofounders will be done well, and it will cost less.

Lesson 3: In selecting your cofounder group, do not worry about the size of the team and sharing of founder stock. Rather, be concerned with building a team of founders who have mutual respect for each other, who are equally passionate about the venture, and who are willing to go through the inevitable crisis with you.

Lesson 4. Take immigration issues seriously, but do not let your immigration status deter you from becoming an entrepreneur. America knows it needs immigrant entrepreneurs. There is always a way for the entrepreneur to stay, even if it costs dearly in legal fees.

CHAPTER 5

Early Funding

Business? It's quite simple. It's other people's money.
—ALEXANDRE DUMAS (1802–1870)

Even if you can afford to work without pay, inevitably you will reach a point when hard work in exchange for sweat equity is not sufficient to keep your company moving forward. You will need to find a source of money that allows you to operate until you either break even or transition toward a more scalable funding model such as venture capital. Which bootstrapping approach is best for you very much depends on the type of tech company you are building.

At ThingMagic we provided design services for cash. We had gotten to know a number of large companies in the context of our graduate student research. When they tried to hire us individually as employees after graduation, we rejected their offers. Instead, we provided them with consulting services as a group, effectively funding ourselves and ThingMagic.

Alternative and creative funding options discussed in this chapter should be considered at later stages of the startup as well, even when venture capital becomes available. Startup teams get excited when they secure large rounds of VC funding. Equity funding *feels* different from a loan, which has to be paid back on a specific date.

In reality, however, venture capital is still borrowed money. In fact, it's the worst of both worlds: venture capitalists want their principal back, and they want a 10x return on top of that. Hence, whenever an alternative funding option is available to your company, go for it (Figure 5.1).

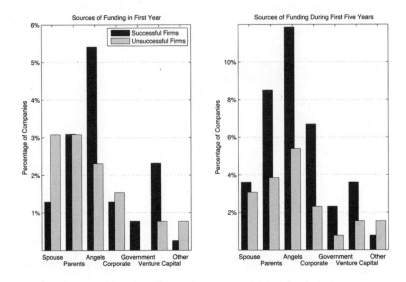

Figure 5.1: Sources of funding of high-tech startups during the first year of operations (left) and within the first five years (right); companies are categorized as *successful* (still in business or sold or exited by 2009) and *unsuccessful* (no longer in business in 2009). Data sources: Alicia Robb and E. J. Reedy, *An Overview of the Kauffman Young Firm Survey, 2004–2009*, 2011, http://www.kauffman.org, and the *Kauffman Firm Survey, 2004–2009*, 2011, http://www.kauffman.org/research-and-policy/kauffman-firm-survey.aspx.

REVENUE, THE BEST SOURCE OF FUNDING THERE IS

Even before ThingMagic had incorporated, an eager technology entrepreneur begged us to design an RFID tag for him. He had neither the expertise nor the staff to do it; we had the skills among our founding team members. But without a bank account, we couldn't even take his check. Instead, he entrusted us with his personal credit card, which we used to buy equipment and materials to fulfill his needs.

Not every tech company is in a position to earn revenue on day 1, but all companies need to switch from burning cash to making money at some point in their life cycle. The sooner you start the process, the better off you are. If you can fund your company with revenue, you don't have to give up ownership and control of the company to investors. You don't waste valuable time raising money from people who know little about your business and whose financial support will end sooner or later. Revenue also doesn't need to be paid back.

So, you ask, "We don't have a product, and yet we are supposed to generate revenue? How are we going to do that?" The first thing to recognize is that your team has specialized knowledge with applicability to certain customers. Granted, your primary objective is to use your knowledge and build a lasting enterprise with a scalable business model. However, in the meantime your knowledge is applicable in other ways as well. Remember, you will never be as smart as you are coming out of school (Chapter 2). You need to capitalize on your smarts and specialized expertise to generate cash while not giving away any IP necessary to build your business (see also the section "Early Customers" in Chapter 3):

- **Technical skills.** If you have outstanding programmers, hardware engineers, or industrial designers on your team, you are in the perfect position to sell those skills to companies who are desperate to get some piece of technology implemented. Try to position your technical services in the context of a particular market or technology, where you can claim expertise. This will help differentiate you from individual contractors and from firms that offer design services as their main line of business. When you sell the time and services of your team, make sure you are not giving up the rights to the very best ideas and implementations that are supposed to be the foundation of your company.

- **Domain-specific consulting services.** As a startup in a particular market, you know your way around your industry. You know who the important players are, what's technically feasible, and how a third-party company could take advantage of the specific market. In a complex market and ecosystem, you will have a few direct competitors and likely a larger number of partners. Turn those partners into consulting customers! Develop a suitable go-to-market strategy for them, and have them pay you for that service.

- **Intellectual property licensing.** Assuming you own intellectual property in the forms of patents, designs, or software, consider licensing your rights in such a way that you do not interfere with the long-term strategy of your company. For example, try selling off the rights to a certain market or a certain geography that is not part of your core go-to-market strategy. This works best if you can pair the IP rights with technical services. In fact, in the best case, you will be paid to develop a platform, which you can then use for your own purposes in the market segment you picked for yourself.

Funding your company with customer revenue is financially highly advantageous. The approach also helps the young startup focus and avoid common pitfalls:

- **Pre-sales slavery.** Resource-rich enterprises rightfully debate endlessly how much they should invest in a particular customer or a particular deal before a contract is signed and the project turns into a revenue-generating activity. In addition, customers tend to exploit the vendor during that period of angst, when the sale is uncertain and the customers have more leverage than they will ever have again. It turns out that expectations on the customers' side decrease significantly when they realize that the vendor is just not in a financial position to provide free pre-sales services.

- **Product pipe dreams.** A small company that is funding itself on customer revenue is unlikely to invest the few available dollars in a product whose value the team is not absolutely sure about. Almost certainly the number of product ideas among a bunch of creative technologists far exceeds the available resources to implement them. Therefore, stakeholders are forced to allocate and put to use the available development and marketing resources very carefully.

- **Feature creep.** Clever customers will tell you that your product is practically perfect and that they will buy a lot from you if only you could provide that one additional feature or "minor change." While the additional work is requested as a prerequisite of doing business together, the customers' willingness to compromise will increase dramatically if you offer to implement their feature wish list in exchange for a development fee rather than for free.

- **No trips to the Bahamas.** A lot of corporate travel is utterly unnecessary but costs a company dearly. For young ventures, the risk of traveling too much is greatly reduced if you have to ask the customers to cover the cost because you have no travel budget. If the customers need you in a face-to-face meeting, they will pay; if they don't, the trip probably wasn't all that important in the first place.

Generating revenue on the side is a most advantageous approach to funding your venture. Unfortunately, it can also be very taxing and strenuous. Effectively, you and your team are doing two jobs at the same time, each of which deserves a full-time commitment. That's not for everybody, and it is hard to sustain this kind of workload for an extended period of time.

Not only are the two tasks competing for the team's attention and time but there is also the constant need to set priorities. Every opportunity, every project, and every task presents a tactical crossroad: Do you work on making a few bucks to fund yourselves, or should you invest time and resources into the future of the company you want to build?

The tension is not limited to engineering and product development. Every corporate function faces a schizophrenic conflict. Business development and sales are challenged to share their time between acquiring projects that generate short-term revenue and building a customer base for the company's long-term product offering. Finance has to constantly evaluate how much is needed to stay afloat short term while not wagering the future of the company in the process.

At ThingMagic, we maintained the revenue funding model for five years. Even beyond the initial bootstrapping period, we would often go back to the original model in order to improve our financials. At later stages, we refined the model to engage in funded projects only when they fit in with the long-term strategic goals of the company. Most important, we pursued funded design projects that involved our very own OEM products. Customers would come to us with the need for a particular RFID end product, which they lacked the resources to develop. We would offer to design a product for them, but we made sure to include our own OEM components in the product. As the projects progressed into volume deployments, our design engagements ended, but we continued to sell hardware components.

INCUBATORS AND ACCELERATORS

Startup incubators had a false start in the late 1990s when a lot of them died in the aftermath of the dot-com bust. A little over a decade later, new incubators, now called accelerators, are being created all over the world. Some are doing exceptionally well, most prominently Y Combinator in California. Many are thriving, including the Tech-Stars network of accelerators with offices in a few entrepreneurial centers across the United States.

In September 2013, Seed-DB counted more than 170 accelerator programs worldwide, more than 2,900 participating or graduated startups, and about $2.7 billion of invested funds. According to the website, 159 exits had occurred, resulting in about $1.7 billion of total exit value.[1]

Interestingly, more than half of these funds were invested by Y Combinator, and more than two-thirds of the exit value was earned by Y Combinator graduates. The fact that the statistics are dominated by a single entity is evidence that the accelerator industry is still developing. Not all of the entities will survive. There are a few spectacular success stories, such as Reddit and Dropbox, all of which graduated from Y Combinator. Time will tell how scalable the model is across regions, industries, and people.

Y Combinator is dominating the accelerator community in terms of its track record. It has also been a role model for other accelerator programs that offer surprisingly similar structures and terms. Participants are paid up to $100,000 in seed funding and possibly receive a larger amount in convertible debt. In exchange, Y Combinator receives 2 to 10 percent of the startup's equity.

Y Combinator holds sessions or classes twice per year. Startups are required to reside in Silicon Valley while participating, but they are free to move anywhere they choose once the session is completed. The three-month-long program is structured around a series of events. On Prototype Day, startups present their current state of development to other participants without the pressure of having outside observers. Halfway through the program, the ventures present to a high-profile venture capital partner. On Rehearsal Day, participants practice their pitch in front of other participants.[2]

Finally, on Demo Day the participants present to a few hundred investors, both angels and venture capitalists, who negotiate Series A term sheets with the companies that piqued their interest.

There appear to be two main reasons for the success of accelerator programs. First, the young startup is pushed to make a lot of progress in a relatively short period of time, stimulated by peers and experienced mentors in an intense environment. Second, the startup has access to a wide investor community on Demo Day.

Conventional fundraising is one of the most time-consuming challenges in the startup experience. It takes a long time to prepare a pitch, get in touch with a potential investor, and then convince him to invest. Investors are picky, so startups typically go through the process many times before they are able to secure funding. The Demo Day concept cuts down the process significantly. With many investors in the room ready to write checks, Series A fundraising turns into a focused exercise of a few weeks rather than a prolonged multi-months drag.

ANGELS

A good friend, who is a successful high-tech executive and spectacularly unsuccessful entrepreneur, confessed to me once that nothing in his short startup endeavor compared to the pain of letting his extended family know that the money they had so willingly invested was gone for good. While one can escape business partners and unrelated investors in the future, one cannot escape family. They will always be there. Risking your relationship with family members for a few dollars is simply not worth it.

Angel investors, on the other hand, know what they are getting into. They consciously invest in risky early-stage ventures, many of which will not make it. In fact, according to the *Angel Investor Survey* undertaken by the Kauffman Foundation for Entrepreneurship, 53 percent of angel investments return less, if any, of the invested money (Figure 5.2).

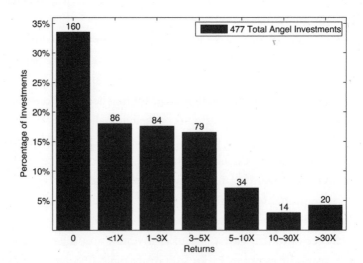

Figure 5.2: Histogram of returns on angel investments in high-tech companies; covers returns for exited or closed investments between 1990 and 2007. Data source: Robert Wiltbank and Warren Boeker, *Returns to Angel Investors in Groups*, Ewing Marion Kauffman Foundation and Angel Capital Education Foundation, *Angel Investor Survey*, Kansas City, MO, November 2007, http://www.kauffman.org.

Angel has become a catchall term for people and institutions that put up equity money for startups but do not qualify as professional investment funds. Traditionally angels have operated on their own, seeking out investments independently. Increasingly, though, individual angels are organizing themselves in larger groups, trying to professionalize the investment process and to leverage each other's time spent evaluating portfolio companies.

Angel investors in technology startups deserve every bit of the positive connotations of the term *angel*. Angels invest when the venture is too small to attract the attention of a venture capitalist. They invest because they know and trust you. They invest because they belong to your field. They invest because they believe the problem you are attacking is worth solving. They invest in your venture because one of their friends did so already. They invest even when financial considerations alone suggest that it would be wiser to pass on the opportunity.

Angels are not philanthropists. As early-stage investors, their risk is greater than that of later-stage VC funds and investments. However, if things work out well, they also experience the biggest gain. In terms of risk and reward, angels are pushing the limits of risk capital.

Many angels have their first career behind them, or they are retired altogether. Most of them have had a good deal of professional success, but they are hungry for more. The Silicon Valley executive who made a few million inevitably would like to be in the Ten-Million-Dollar-Plus Club. The entrepreneur who made tens of millions on a successful exit would rather have $100 million in his portfolio. And so on. Angel investing offers the possibility of increasing one's net worth without working a 24/7 job.

Smallish angel rounds are typically structured as convertible debt. Since it is hard to put a valuation on a venture that has nothing to show for itself, the valuation question is postponed until such time as a more significant round is being raised. When the debt converts in the future, the early angel's investment is converted into equity at a discount of the round's stock price.

Angels invest for a variety of reasons, many of which are not financial. Before you approach specific angels, do your homework. Find out what motivates them, what kinds of companies they like, and how they can help you be successful beyond providing funds. The more you can leverage their talents and preferences, the easier it will be to convince them to fund you and the more you will get out of the relationship.

FREE MONEY, COURTESY OF YOUR GOVERNMENT

The U.S. government is not known for generously giving away money in support of for-profit endeavors. And yet, the government is one of the few places in America where you can get free funding for your commercial technology venture. Money from government grants does not carry any interest, it does not have to be paid back, and you do not have to trade equity to get it.

The Valley of Death

Early-stage technologies hit a particularly vulnerable phase in their life cycle in between university-based research and commercialization. When a technology has been far enough developed to no longer interest the academic community, but it is not sufficiently far along to represent a manageable commercial risk, profit-motivated financial backers are hard to find. This transition period has come to be known as the Valley of Death. When neither venture capitalists, nor angel investors, nor large corporations are willing to expose themselves to the financial risk of the new technology, the technical fledgling likely dies before it gets a chance to make the world a better place.

The Small Business Innovative Research (SBIR) program was established by Congress in 1982 to address the Valley of Death problem. The program was chartered to support the research and development activities of small and medium-size businesses and to help promising, early-stage technologies through the Valley of Death.

In 1992 the Small Business Technology Transfer (STTR) program was established as a sister program to the SBIR program. The STTR grants support the collaboration between small technology businesses and not-for-profit research organizations (Figure 5.3).

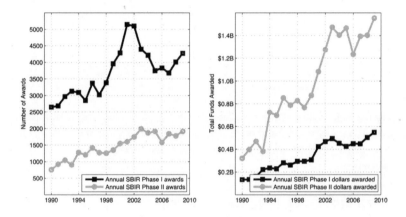

Figure 5.3: Annual SBIR awards—all agencies, all U.S. SBIR awards (left) and all agencies, all U.S. SBIR dollars awarded (right). Source: *SBIR Awards by Agency and Year, 1983–2012,* www.sbir.gov/past-awards.

The goals of the two programs are multifold:

- Stimulate technological innovation

- Meet federal research and development needs

- Foster and encourage participation in innovation and entrepreneurship by socially and economically disadvantaged people. (Figure 5.4)

- Foster technology transfer through cooperative R&D between small businesses and research institutions

- Increase private-sector commercialization of innovations derived from federal research and development funding[3]

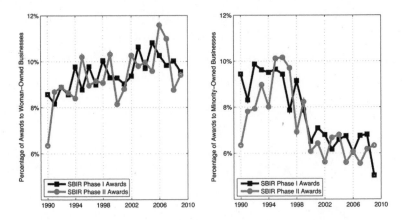

Figure 5.4: Annual SBIR awards for 2010—percentage of awards to woman-owned businesses (left) and percentage of awards to minority-owned businesses (right). Source: *SBIR Awards by Agency and Year, 1983–2012*, www.sbir.gov /past-awards.

Small Business Innovative Research

The SBIR and STTR programs are administered by the individual award granting agencies: the National Science Foundation (NSF), the Department of Defense (DOD), the National Aeronautics and Space Administration (NASA), the National Institutes of Health (NIH), and the Department of Energy (DOE), among others (see Figure 5.5). Each agency has implemented its own processes, including solicitations, submission deadlines, and program directors. Companies apply directly to each agency.[4]

The SBIR program offers two main funding phases: Phase I funding is limited to $100,000 to $150,000 in grant money and six months of research. Funding is provided to establish the technical merit, feasibility, and commercial potential of the proposed R&D effort. Furthermore, the recipient of a Phase I grant is given an opportunity to prove itself capable of complying with the general guidelines of the program.

Upon completion of Phase I, the recipient can apply for a Phase II award, which consists of up to $1 million in funding for a duration of two years. The Phase II awardees are selected based on the Phase

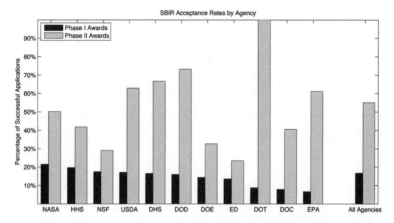

Figure 5.5: SBIR acceptance rates by government agency for 2010—Phase I and Phase II awards. Source: *SBIR Acceptance Rates by Government Agency, 1983–2012,* www.sbir.gov/past-awards.

I results as well as the specific work proposed for Phase II. The commercial potential of the technology becomes an important selection criterion for Phase II awards.

Some agencies—notably the Department of Health and Human Services (DHH) and the NIH—exceed the above funding limits for Phases I and II. Also, some agencies offer follow-up funding through other programs and contracts, mostly with the intent to satisfy a particular governmental need for a product or service (Phase III).

Participating businesses have to meet the following criteria:

- The business must be incorporated as a for-profit corporation in the United States.

- The business must be at least 51 percent owned and controlled by citizens or permanent residents of the United States.

- The business, including its affiliates, cannot have more than 500 employees.

- For SBIR applications, the principal investigator must have his primary place of employment with the company.

- For STTR applications, the for-profit small business corporation must partner with a not-for-profit research organization to perform the project.

Importantly, you do not have to be incorporated when you apply for a grant. However, you need to incorporate a for-profit company at the time that you accept the grant.

SBIR and STTR Strategies—for the Love of Technology and Innovation

The SBIR or STTR application processes are NOT business plan contests. The purpose of the program is to fund innovative research, not to commercialize such innovations. While the commercial applicability is an important consideration in the funding of a project, the award money is not provided for product development or commercialization per se.

For some small technology companies, SBIR grants are a primary source of revenue. The business model in which you move from grant to grant won't make you rich, but it is rewarding if you like conducting research and would like to stay in charge of your destiny. However, be aware that the program is not intended as a permanent funding vehicle. You should expect your chances of acceptance to decrease after the completion of a few grant projects.

Since there is no guarantee that a proposal will be accepted, you have to apply for more grants than you can actually deliver on. In 2010, the average SBIR acceptance rate was 17 percent for Phase I proposals and 55 percent for Phase II proposals (see Figure 5.4). The acceptance rates for STTR proposals are slightly higher at 22 percent for Phase I proposals and 68 percent for Phase II proposals.

The SBIR and STTR application processes are inherently anonymous. Indeed, you can apply for a grant without ever communicating with any program manager in person. However, your chances of acceptance increase dramatically if you build personal relationships with the program directors working in your field. SBIR program directors have two responsibilities: (a) coming up with new research topics and solicitations and (b) finding suitable companies to work on the topics. To do a good job with either task, program directors rely on the input and contributions of subject matter experts. They need a relationship with you, and you need a relationship with them. As with most proposals, you have the best chance of acceptance if you provide input into the solicitation yourself.

You will be the owner of all intellectual property generated during a funded project. However, the government has the right to use the IP in any way it pleases. This policy is often seen as a disadvantage of SBIR funding, but in practice, the sharing of rights has little negative impact on the future monetization of the IP. If you are lucky enough that some agency wants to use your work product, the agency will most likely ask you, the technical expert, to commercialize the research. This allows you to capitalize on your work and get paid by the government a second time. If, on the other hand, the government is not interested, you can sell the product elsewhere.*

If you need money desperately and quickly, the timing of the SBIR application and award process may not work for you. The application process starts a couple of times each year. However, it takes a few months from the application deadline for businesses to be notified about their acceptance or rejection, and it takes time for the funds to be provided. For example, the NSF has two grant deadlines per year, and money is provided only about six months after each application deadline.

Also, Phase II is where the money is. Don't waste your time on a proposal that you do not intend to take beyond Phase I. Even if you get Phase I funding, you won't be able to stay financially ahead on $100,000. Only Phase II awards provide enough money to execute the project with positive cash flow and to accomplish serious research and development.

---

* That said, it is advisable to consult with a government contract lawyer. There can be IP-related traps when contracting with the government.

83

BANKS AND LOANS

The U.S. venture funding industry is based primarily on the concept of equity financing. And yet, when we sold ThingMagic, the bank that had provided a credit line to the company booked a bigger gain than any of our venture capital investors. How is that possible?

Venture Lenders: If *They* Believe in You, So Do We

Banks that specialize in venture lending follow venture capital. Rather than figuring out the viability of a certain venture themselves, these venture lenders co-invest along with or immediately following an equity round. The banks make the assumption that a company in which venture capitalists invest must be at least moderately stable. The banks also consider the viability of the VC firm itself, knowing that a financially sound investor will inject more money into needy portfolio companies down the road.

Venture lenders know how to establish a healthy upside for the credit they are giving to startups, while protecting themselves on the downside. When a venture fails, equity investors may not get their money back. However, both startup executives and VC partners move heaven and earth to pay back outstanding bank loans, so as to not stigmatize themselves with a default. Nobody likes to damage their reputation with the banks, even if that means an additional short-term loss.

Under which terms do banks lend to startups in a co-investment?

1. The credit is provided in conjunction with an equity round.

2. The credit is considered the most senior debt in the company.

3. The size of the bank loan is smaller than the money invested by the venture capitalists.

4. The bank loan is redeemable and needs to be paid back irrespective of how well the startup is doing when the principal comes due.

Given these rather tight rules, a company would have to go bankrupt and be worth literally nothing for the bank to lose all its money. Even if the company came to such a disastrous end, the alarm bells would go off way before, leaving the bank with time to pursue various strategies to get its money back.

On the upside, banks have a number of tools to make the credit worth their while. First, they charge interest rates significantly above those applied to regular commercial or personal loans. Second, the bank expects to be issued warrants. Last, the deals tend to include a success fee in case the venture capital funders exit during the lifetime of the loan. Hence, the bank is in a position to maximize its return on investment, even though other investors and stockholders might be losing money in the exit.

From the point of view of the startup, this kind of co-investment unfortunately does not eliminate the need to secure equity funding first. Rather, the bank credit is an instrument that can help you embellish and stretch the money you receive from equity investors. Before you consider such a loan, pay close attention to the next section.

Watch Out for the Covenants!

Bank loans destined for startups come with a catch: covenants! Banks like to establish performance benchmarks for the company to further reduce the risk of default. The startup loan is considered in good standing as long as the company maintains a certain level of revenue, a well-defined bottom-line performance, and a minimum cash balance.*

Let's consider the following example: In the process of negotiating a venture loan, you suggest that the company will continue to grow its revenue at a rate of at least 20 percent and that any quarterly loss will not exceed $200,000. The bank manager expresses his trust in your management skills, and he says he wants to add some margin in case your plan does not work out exactly as hoped. He establishes a covenant to call the loan in the event that your revenue decreases or

---

* In addition to financial covenants, venture debt deals typically include a "material adverse change" covenant, which is less well defined but equally disconcerting for the venture.

the cash balance drops below $500,000. You happily agree because from your point of view, it is utterly unimaginable that the company might ever do as badly as that.

Only six months later, the economy takes a turn for the worse, completely destroying the growth prospects of your startup and reducing your cash balance way below the $500,000 limit.

When covenants are tripped, the bank has the right to foreclose, and in extreme situations it will make use of that right.* However, banks usually prefer to renegotiate rather than foreclose the business right away. Foreclosures are messy and hard work for the bank, while rarely resulting in the recovery of the principal. Instead, bank managers and startup managers negotiate a new set of terms that hopefully will lead to the survival of the company and the repayment of the principal.

There is great irony in the fact that the very cash balance a company tries to embellish by getting a loan or credit line with a bank is used as a metric to possibly call that loan. In practice, this means that only a portion of the loan amount becomes true operating capital. You need to hold on to some of the money so as not to trigger the covenant. And you have to pay interest on that money, which is sitting in your own bank account!

Personally Guaranteed Loans

Banks may be more willing to lend money if the founders or executives are prepared to personally guarantee the loan or put up some personal collateral. However, this approach is problematic in so many ways:

- **"How do I explain this to my spouse? Now or when the money is lost?"** Providing a personal guarantee puts enormous stress on a founder's personal finances. Even if the risk is manageable, just having to explain the guarantee to one's spouse can be daunting.

---

* If the bank forecloses the business's loan, it will typically proceed with a fire sale. Since the employees are mostly gone at that point, the value of the enterprise is reduced to its fixed assets and possibly some IP. Not a pretty or lucrative scenario for anyone involved, including the bank!

- **"Why do I have to pay an extra high interest rate?"** Even though the loan has a personal guarantee attached to it, it is still a commercial loan. The interest rate for a commercial loan is significantly higher than it is for a personal loan such as a mortgage on a house.

- **"Why should one of us be singled out for financial sacrifices?"** The founders and major stakeholders in a startup often have different financial backgrounds. Getting all of them to participate in a guarantee is almost impossible to do. Yet, a guarantee in which only one founder or a subset of founders participates can create tension and divisions among the founders.

If the company truly needs one or more of its founders to pony up some cash, it makes more sense for the founders or stockholders to loan the money to the company outright. In practice, this means that one of the founders either taps into his cash reserves or takes out a home equity line of credit. The founder then turns around and loans the funds to the company under well-defined terms.

The founder should be doubly rewarded for the risk taken: the venture should pay an interest rate consistent with commercial rates or even higher, so that the creditor-founder benefits from the interest rate spread between his own loan and the loan granted to the company. Furthermore, the company should grant additional warrants or stock options to the founder lending the money. After all, he is really acting as a venture lender.

CROWDFUNDING

This section would not exist if this book had been written only a couple of years earlier, even though crowdfunding has been a century-old method of financing. The pedestal for the Statue of Liberty was one of the early projects made possible by a crowdfunding campaign. In 1884, newspaper publisher Joseph Pulitzer raised over $100,000 from more than 125,000 people to complete the pedestal, which, unlike the statue itself, was not provided by the French.[5] Crowdfunding has more recently reached sudden popularity due to a few key events:

- In 2012 Congress passed the **Jumpstart Our Business Startups (JOBS) Act**, according to which crowdfunding is set to become a legal method of fundraising.

- A number of **crowdfunding websites** facilitate the process of matching up investors and companies seeking money.

In the past, crowdfunding has mostly been used in support of creative projects, for example, a piece of art, a concert or event, a social cause, or . . . to put a big statue on a pedestal. As no securities were sold, none of these projects qualified as investment activity, and hence the legal framework was very simple.

When crowdfunding is used for startup funding and securities are being issued, the funding event counts as investment activity, and hence a whole set of different rules apply. The JOBS Act is meant to make this easier and to allow companies to raise a limited amount from individual—nonaccredited—investors. Unfortunately, it will be a little while longer before Congress establishes clear rules on what is allowed and what is not. So far, two types of tech crowdfunding have emerged:

- **Why not invest in what you believe in *and* make a big return?** Equity-based crowdfunding provides investors with a piece of equity in exchange for cash, in much the same way that traditional equity financings are set up.*

- **We *will* make it if you pay up.** Reward-based crowdfunding promises to give the investors a gadget, product, or service at some time in the future in exchange for money paid today.

The second option much resembles a futures contract. Two business partners agree to a business transaction in the future, but they

---

* At the time of this writing this is not yet legal, but regulations are expected to be issued.

settle the deal today. The buyer is exposed to the risk that he will never see the product and he will not get his money back if things don't work out.

Startup crowdfunding is useful for some business models and products but not others. Since crowdfunding fundamentally addresses a consumer audience, business-to-business (B2B) startups are unlikely to benefit from this tool anytime soon.

Kickstarter, one of the most successful crowdfunding sites to date,[6] requires applicants to have some sort of prototype. The applicant needs to specify his funding goal and then reach that goal in order for funding to be released. Projects regularly exceed their goal manyfold, which is permissible.

The applicant needs to realize that he is not getting money for free and that he is expected to make good on his promise. Especially when hardware is involved, the economics can get tricky. Assume that you are selling to enthusiastic patrons for $100 a gadget that is not fully developed. Assume further that you are able to convince as many as 10,000 people to buy into the proposition. Excited, you go about spending the million dollars you raised just to complete the product development and design for manufacturing. When you are all done, you realize that the new and fabulous gizmo actually costs $70 to manufacture. You now have to find an additional $700,000 to manufacture the devices before you can satisfy your crowdfunding supporters.

No serious entrepreneur would forget about the difference between revenue and gross margin, or would he? As a matter of fact, even the best of us have managed to underestimate the cost of goods of a new product. In crowdfunded product development, such bad planning can mean the end of a gadget dream and a ruined reputation.

On the positive side: a well-run crowdfunding campaign not only brings in funding but essentially kicks off the marketing and sales activities of the venture. Without doubt, crowdfunding is the sexiest funding option there is at the moment. The coming years will show whether it has a lasting impact on high-tech venture funding.

CREATIVE MONEY

Lesson 1: Early revenue from customers is the most desirable source of startup funding for so many reasons.

Lesson 2: Participation in an accelerator program can help you make a lot of progress toward building your venture in a short amount of time, and it will help streamline the VC funding process.

Lesson 3: Remember that angels are not just investing for monetary gain. Your angel investors become true assets when you leverage their connections, experience, and—in some cases—vanity and let them contribute where they can be most helpful.

Lesson 4: Regularly check R&D grant solicitations to see if you can get free funding to develop your technology offering.

Lesson 5: In the U.S. startup world, bank loans are a short-term financial gap filler, rather than a long-term financing instrument. Stay away from personally guaranteed loans, and make sure to carefully read the covenants of any credit line you take out!

Lesson 6: Crowdfunding is the sexiest of them all! At its best, crowdfunding will get you money to do what you want and a ton of publicity along with it.

CHAPTER 6

Administration

Some rise by sin, and some by virtue fall.
—William Shakespeare (1564–1616)
Measure for Measure

Starting a technology company requires a great deal of administration. Some of these tasks look hard but are easy; some are just easy. No administrative chore compares to the daring objective of inventing good technology and selling it successfully. Use this chapter as an administrative checklist and guide to get the boring stuff out of the way.

LEGAL REPRESENTATION

A company's incorporation is commonly handled by a lawyer, which means you should look for legal representation before your venture even exists. Most law firms are going to be eager to work with new clients, so don't be shy about cold-calling a few firms and comparing their offerings.

Attorneys usually don't charge you while they are trying to win you as a new client. Hence exploratory calls are the cheapest interaction you will ever have with your law firm. Never again will you have as much leverage to negotiate than prior to becoming their client. You could ask for a flat fee for your company's incorporation and other startup services. Or, you could make the payment of fees contingent on your getting your first round of funding. Any of these requests is perfectly reasonable.

In selecting a firm, consider the following issues.

Chemistry

Make sure that you are comfortable with the firm and that you get along well with the partner you will be working with on a day-to-day basis. I fondly remember ThingMagic's incorporation meeting. I may not have grasped the implication of all the issues our attorney raised, but I felt completely at ease with the way he handled the discussion.

Affordability

Even if you consider the notion of an affordable law firm an oxymoron, try to contain the damage. Consider the overall fee structure of the firm and the specific rates for partners as well as associates. Surprisingly, legal rates do not vary all that much by region. However, they do vary a great deal by firm size. On average, firms employing more than 150 lawyers charge about three times as much for partners and associates compared to firms with fewer than 9 lawyers.[1] Going small can save you big-time!

Firms will argue that they will work with you to keep fees low by assigning junior associates to jobs whenever possible. In my experience, the involvement of associates is the surest way to blow one's legal budget. Junior associates have little experience, and most would admit that law school didn't exactly prepare them well for real life as an attorney. Furthermore, associates are expected to run up a minimum of billable hours, which is not the best motivator to finish a job quickly.

Availability

Find a firm with an attorney team that is available when you need help. You need your lawyer to do work for you on short notice, and you can't afford having to bring another attorney up to speed just because your main guy is busy on another deal.

Large firms with national and international offices like to argue that they can represent you wherever you need legal help. This happens to be an utterly useless benefit for a small startup. You will most likely have one office location, preferably in the same city as your law firm. You are also not planning on becoming involved in a lawsuit any time soon, let alone in another state or country.

Services

Your primary law firm should offer enough of a range of services to get you through the first few months of operations, including corporate law, tax law, employment law, immigration law, venture capital law, securities law, financing law, copyright law, and trademark law. If these areas are covered, your law firm will be sufficient for the foreseeable future.

Large firms advertise themselves on the notion that they offer a full range of services. They cover every aspect of law including litigation, patents, and international law. In reality, "full service" means you get to work with specialist attorneys inside the big firm, and you pay big bucks for those services.

Instead, pick a small firm with a generalist business attorney who has a good understanding of your business and the context in which you are operating. In the rare event that you do need a specialist, you can find someone in another firm who truly knows what she is doing. You wouldn't expect your primary care physician to handle all your medical problems between now and when you die either.

Patent Work

Patent law and patent filing expertise is a nice-to-have specialty within your primary law firm, but it is not a requirement. More likely, you will engage with an expert patent attorney in your field or with someone you know already from previous patent work. You would constrain yourself unnecessarily if you stuck with your general business law practice for patent work.

Also, IP lawyers are particularly skilled in extracting extraordinary amounts of money from you when you're not paying attention. Try to maintain relationships with more than one firm so that you have leverage each time you do a new filing.

Once you are working with a firm, always limit the time and fees on a project beforehand. Also, never be shy about asking for a reduction in fees after the fact. Large companies do it all the time, especially when the work product didn't have the desired effect—for example when a deal fell through. Attorneys are supposed to keep you in business, not ruin you financially. You just have to remind them of that little detail once in a while, lest they forget.

You will almost certainly change law firms during the lifetime of your company. There is no need to hire the best and most pricy firm from the start. Unfortunately, even the priciest firm can't make your company successful. You will have to do that yourself.

CORPORATE FORM

Choosing a corporate structure may feel like a boring and secondary issue to anyone with a great idea and the determination to found a company. After all, how many options could there be? Indeed, there aren't many options to choose from, but the differences between them are significant. Choosing the right legal form can contribute greatly to a smooth operation and to financial reward. The choice of corporate structure also sets the strategic tone with regard to venture capital investments, management earnings, and governance (Figure 6.1).

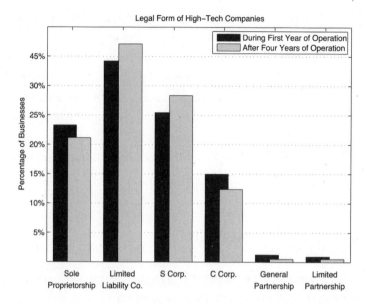

Figure 6.1: Legal form of high-tech startups in year 1 of operations and after four years of operations. From Alicia Robb and E. J. Reedy, *An Overview of the Kauffman Young Firm Survey, 2004–2009*, 2011, http://www.kauffman.org, and the *Kauffman Firm Survey, 2004–2009*, 2011, http://www.kauffman.org/research-and-policy/kauffman-firm-survey.aspx.

Regardless of which legal form you pick, you have the option of incorporating in your home state or in Delaware.* While more than 63 percent of U.S. Fortune 500 companies are incorporated in the state of Delaware,[2] only one of them is actually headquartered there (DuPont[3]). Corporations choose Delaware as their legal home for a number of good reasons, which apply equally to large and small entities:

- The Delaware general corporation law (DGCL) offers advanced and mature corporate laws for U.S. businesses.

- The Delaware business court has a history of more than 200 years generating business case law clarifying the DGCL.

---

* Technically you can incorporate in any other state as well, but such a move makes sense only in very particular circumstances.

As a practical matter, you will be doing business with companies all across the country. You will find that many of those business partners are incorporated in Delaware as well. As you draft agreements and engage in contracts, both parties will try to make the law in their home state apply, but most companies will be willing to compromise on Delaware law. Furthermore, most venture capital funds require their portfolio companies to be incorporated in Delaware. Incorporating in Delaware costs you slightly more in fees, but the extra expense is well worth it.

Limited Liability Company (LLC)

The limited liability company (LLC) is the most flexible of all U.S. corporate forms. You choose the legal form of LLC in order to take advantage of the LLC's favorable tax treatment or to ensure that you have flexibility around ownership and governance policies. LLCs are subject to few regulations; hence you can create a corporate structure to your liking as far as ownership structure, capitalization table (cap table), and officers and governance are concerned. You can make an LLC look like a corporation, or you can make it look like a classic partnership, and you can define a hybrid between the two.

The defining property of LLCs stems from its treatment by the U.S. tax law. LLCs are subject to pass-through taxation. The entity itself does not pay income taxes, but the tax liability flows through to the personal tax form of the individual member of the LLC. While this may sound bad, it is actually the key to saving big-time on corporate and personal tax payments.

Unfortunately, this flow-through characteristic is one of the reasons why LLCs are not very popular among certain stakeholders, notably investors. Venture capitalists do not like to invest in LLCs and in many cases are prohibited from doing so. If your business requires significant third-party capital, this is a serious concern, and you should consider incorporation as an S or C corporation instead.

LLC Ownership Structure and Governance: Anything Is Possible. LLCs are owned and controlled by their members. The governance rules are defined in the LLC's operating agreement, which constitutes a contract between the members. The level of the individual member's ownership is manifested in the number of membership interests she owns.

Voting rights among the members can be defined in almost any imaginable way. A decision may require a simple majority of the members or unanimity or anything in between. For example, the operating agreement can require that a supermajority of members agree on far-reaching financial decisions or a decision to remove a member but that only a simple majority agree on day-to-day matters.

The voting members of the LLC elect one or more *managers*. The managers act like the board of directors in a corporation. In addition, an employee manager frequently holds the executive power of the LLC and runs the day-to-day business. She can designate any number of additional executives, including a CEO. Those executives can be given any titles, and they can be given any amount of authority and executive power. The additional executives may or may not be managers or members themselves. The LLC is not required to nominate a president, treasurer, and secretary as with a corporation. There are no formal requirements for the members to meet on a regular basis, but the operating agreement can suggest that such regular and mandatory meetings take place.

Additional classes of members can be defined, such as nonvoting members. A nonvoting member benefits from profits and distributions of the LLC. However, the nonvoting member has no say in whether or not that distribution is made in the first place, nor does she get to vote on any other company matter.

A stock option plan can be established for an LLC, but the process is more complicated than it is for C corporations. You will have to define the plan from scratch. Furthermore, there is no such thing as *incentive stock options* (ISOs). While you can create an option plan in an LLC, the options do not get the same beneficial tax treatment as ISOs in C corporations (see the "Equity" section at the beginning of Chapter 7). This can be a serious issue in attracting key employees who expect simple and familiar stock options and restricted stock, not member units.

Dealing with the departure of a member is one of the more delicate and tricky issues in LLCs, even if the parting of ways is amicable. The operating agreement needs to specify in detail how much ownership a departing member can take with her. In a closely held lifestyle company with just a few owners, there often is a stock buyback clause for the company upon the departure of a member. In a high-tech startup groomed for an exit, on the other hand, employees rightfully expect to be able to take stock with them when they leave.

Severing a member against her will can easily turn into an emotional and legal nightmare. The founding team better make sure that there are good rules in place for this unfortunate situation. The more carefully you drafted your operating agreement in the first place, the less painful the negotiations will be in an actual forced separation and the better are the chances that you can stay friends despite the professional disagreement. When you design these rules with your cofounders, keep in mind that you could be the first to be kicked out.

LLC Pass-Through Taxing: Great if You Intend to Make Money. LLCs don't pay taxes on profits earned. Rather, the tax liability flows through to the individual members. In simplest terms, this usually means that the profits of the LLC are shared among the members according to their pro rata ownership. Before the annual tax filing date, the company files a tax return and reports each member's share of annual income as well as guaranteed payments on K-1 forms. The member then pays income taxes on the reported profit.

In any given year or quarter, one of the following scenarios takes place:

- **Distributed income.** The company has had net income (profits) and decides to distribute the income to its members. Individual members pay income tax on the money received.

- **Retained income.** The company has had net income (profits) and keeps the earned cash in the company as operating capital. In this case, each individual member pays income tax on the retained income, based on the pro rata share of her ownership. Since members are asked to pay taxes on money they did not receive, retained earnings can cause controversy between LLC executives and members. If the company can afford it, it should pay the individual enough in cash each quarter so that she can cover her personal tax liability. Frequently this is made a requirement in the operating agreement.

- **Losses.** The company has negative income in a given quarter or year. It reports the loss on the K-1 forms for the individual members as a loss equal to her pro rata ownership. As such the pro rata loss offsets other self-employment income of the individual.

- **Guaranteed payments.** The company makes guaranteed payments to some of its members in return for services rendered. Like partnership income and loss, guaranteed payments are reported on the individual's K-1. Unlike regular salaries and wages, guaranteed payments are not subject to withholding. The recipient pays ordinary income and self-employment taxes on the guaranteed payments.

Given these basic scenarios, members and member-employees of LLCs enjoy significant tax-related benefits compared to stockholders in C corporations. Most important, a dollar earned by an LLC is not being taxed twice as is the case for dividend payments of C corporations. Furthermore, since the pass-through income is treated as self-employment income, unreimbursed business expenses can be deducted directly by the individual.

Lastly, the company may qualify for certain tax credits. In a corporation, tax credits can be applied only against the income of the corporation. In an LLC, tax credits apply to the member's tax return and hence reduce the individual's tax obligation.

C Corporation

A U.S.-based C corporation (C corp) is controlled by its stockholders who own one or more classes of stock as defined in the certificate of incorporation of the company. Classes of stock may vary in terms of voting rights, and they offer different degrees of liquidation preference. Founders, executives, and employees are typically awarded common stock, while investors typically receive *preferred* stock.

A board of directors is elected by the stockholders and is responsible for managing the affairs of the company. Day-to-day control of the C corp is delegated to the officers of the corporation who are elected by the board of directors.

The rules by which the C corporation is governed are manifested in the *bylaws* of the corporation. The bylaws determine the authority given to the executive team and the board of directors. The bylaws also specify how the board is composed and elected and how often board and stockholders meet.

The C corporation must nominate a president, a treasurer, and a secretary, but all three positions can be held by the same person. The officers of the corporation, including the president and the CEO* make the day-to-day business decisions for the corporation. Important financial and strategic decisions have to be sanctioned by the board and in certain cases by a stockholder vote.

The board meetings are run by the chairperson, assisted by the secretary. Board resolutions are best shared in writing before the actual board meeting, so that potential issues can be addressed ahead of time. If the board meeting turns into a fight, the executives haven't done their homework preparing it.

Private C corporations are not required to hold in-person stockholder meetings. Rather, stockholder resolutions can be adopted when a majority of the stockholders provide their written approval. The company is required, however, to notify all stockholders after the fact that a resolution was passed.

---

* Note that there is no requirement to name a CEO.

Most instruments of governance in C corporations have been invented for a reason. However, unless you go public or are sponsored by unfriendly investors, many regulator-imposed rules on how to run your company can be reduced to formalities if you so choose.

For example, it is technically sufficient to name a single member to the board of directors. Assuming you make yourself the sole director of your board, your life just got a lot easier. However, directors can actually be helpful in running your business. You may not always like what they have to say, but you might as well take their input seriously. Likewise, the prospect of facing angry stockholders at the annual stockholder meeting may seem daunting, but these meetings actually offer a great opportunity to reflect on the past year and communicate with your employees and other stockholders.

Note that all of the three legal forms discussed in this chapter are limited liability companies, which means stockholders are not liable beyond the value of the stock they own. Should the company go bankrupt, the stockholders are not responsible for the losses of other creditors such as banks and suppliers. In other words, you are not going to lose your house if you—or the CEO you hired—mismanaged your company. This is true for stockholders, but not strictly speaking for directors and officers. For example, if the company is unable to make payroll, the officers and directors of the company are liable for any outstanding salary and wage payments.

S Corporation

The S corporation (S corp) is best understood as a hybrid between a C corporation and an LLC. While the S corp implements the same corporate structure and governance principles of a C corporation, its tax obligations follow the model of the LLC. Consequently, S corporations do not offer as many governing options as an LLC, but they are easy to understand and quick to set up. S corps are required to appoint officers, hold regular board meetings, and schedule annual stockholder meetings.

Like LLCs, S corps are not subject to corporate taxes. Instead, they pass all earnings on to the stockholders, who in turn pay income taxes. Like LLC members, S corp stockholders get angry when they get hit with a tax bill for retained earnings.

To qualify for S corporation status, the company must submit Form 2553 (Election by a Small Business Corporation) signed by all the stockholders, and it must meet the following requirements:

- The company must be a domestic corporation.

- The company can have only certain types of stockholders: individuals, certain trusts, and estate stockholders are allowed. Partnerships, corporations, or nonresident alien stockholders are not allowed.

- The company can have only up to 100 stockholders.

- There can be only one class of stock.

- Certain types of corporations are not eligible, including certain financial institutions and insurance companies.

Incorporation as an S corp is a good choice in the following situations:

- **You want your startup to look like a C corp but operate like an LLC.** You need your startup to look like a C corporation, but you want to enjoy the tax benefits of pass-through tax laws. Certain constituencies do not trust the LLC structure. Investors, in particular, like to know what kind of entity they are dealing with, but LLCs can be enigmatic. S corps, on the other hand, are governed like C corps, and hence they can be well understood without having to read through 50 pages of an operating agreement.

- **You need simplicity.** You would like to establish a simple structure for your startup that does not require the *manual* definition of every single detail. Sole proprietors especially do not need the degrees of freedom and overhead that an LLC offers. Incorporation as an S corp gets you going quickly.

- **You might convert to a C corp in the future.** You want the option to convert your business to a C corporation at a later time. While it is possible to convert both an S corporation and an LLC into a C corporation, the mechanics of the process are much easier for the former. You simply send a letter to the IRS notifying them of the revocation of S corporation status. It should be noted that the reverse is not true. Once you are incorporated as a C corp, it is very difficult to convert to an S corp or an LLC.

ThingMagic started as an LLC, which was a good choice for our early years when we made money by providing design services to others. When we turned ourselves into a product company and started to look for venture capital, we converted the company to a C corp. Even though the company was still reasonably simple, the legal work to implement the conversion seemed endless. The conversion process, followed by the venture capital negotiation, also brought many of the emotions and anxieties of the initial company founding back: How much stock does each of us get? What will our titles be? A new corporate form means many existential questions have to be answered again.

CORPORATE COMPLIANCE

Your corporate identity is established by means of a few key documents. Some have to be in place and signed by the time of incorporation, others need to follow soon thereafter, and most will be amended over the lifetime of the company.

The Charter

The charter—also called the articles of organization, certificate of incorporation (corporation), or certificate of formation (LLC)—lists the principal place of business of the corporation, its purpose and the type, and number of shares of stock issued. The charter is filed with the state government, and it usually carries a filing fee. Delaware corporations file their charter in Delaware and must also qualify to do business in the state of their principal place of business. Delaware corporations also pay incorporation and annual fees in both states.

The Corporate Bylaws

The corporate bylaws (corporation) or operating agreement (LLC) are created at the outset of the company and complement the charter. The bylaws specify the details of corporate governance and operations of the corporation. Provisions include the composition of the board of directors, the number and type of officers in the company, the procedure of electing executives and directors, stockholder voting rights, and procedures for board and stockholder meetings.

The filing of the charter constitutes the formal formation of the company. The bylaws are not required at the time of incorporation, but they should be signed soon thereafter to avoid misunderstandings among the founders or other early contributors.

The Stockholder Agreement

The stockholder agreement specifies any special provisions relating to voting, rights and obligations of the individual stockholders, restrictions on transfer, and so on. In the case of an LLC, the stockholder agreement is usually part of the operating agreement.

Stock Certificates

Stock certificates are to be issued in connection with every stock award to a founder, employee, or investor. Stock certificates are usually printed on special company stock, which you can custom order and then print with a regular printer.

83(b) Elections

83(b) elections ensure beneficial tax treatment of restricted stock awards (see the "Equity" section at the beginning of Chapter 7). The election is made by the individual stockholders within 30 days of the issuance of restricted stock. The election is necessary only in cases in which stock can be repurchased—that is, if the stock grant is subject to vesting.

Tax Forms

Tax forms and tax filing requirements vary from state to state (see the section "Bookkeeping and Taxes" later in this chapter). You should assume, though, that you are obligated to file tax-related documents very soon after incorporation to stay in compliance. States tend to maintain comprehensive websites that outline the specific requirements. After all, that's how they make money.

The Minute Book

The minute book holds copies of all official documents relating to the incorporation, the bylaws, stockholder meetings, board of director meetings, and other compliance issues. It is a repository of every important document relating to the governance of the company. A good laundry list of documents, events, and decisions that need to be included in the minute book helps a great deal in keeping the company in compliance:

- Charter and any amendments

- Bylaws and any amendments

- Approved minutes of board meetings (Minutes taken at one meeting, usually by the secretary, are approved by the directors at the next board meeting.)

- Board resolutions

- Minutes of stockholder meetings

- Stock and option agreements

- Stock certificates (unless they are given to the stockholder)

- Option certificates

The minute book is kept either at your primary law firm or on your premises.

Corporate compliance is rather inexpensive compared to many other regulatory requirements specific to your business and industry. Have a look at the following list of compliance areas and determine which ones you should be worried about:

- Electromagnetic compliance (Federal Communications Commission [FCC] certification in the United States; similar regulations in other parts of the world)

- Federal Department of Agriculture (FDA) approval

- Lead compliance (mandatory compliance for electronics sold in Europe; de facto required anywhere)

- Certificate of occupancy for manufacturing and lab environments

- Kosher compliance (if you are in the food industry)

- Export compliance (You can't ship to a small number of countries [the infamous *axis of evil*] at all, but that's only one rule out of many.)

- Hazardous or international shipping (who would have thought that shipping could be that hard)

However long your list, don't get discouraged! Remember that your competition is subject to the same rules and that you are smarter and better than they are. If you have your own house in order, you are in a stronger position to exploit the compliance weaknesses of your competitors. Customers hate nothing more than being dragged into a messy compliance situation.

A few years ago one of ThingMagic's competitors *decided* to ship radios without FCC certification. Someone (not us, mind you) told the FCC and the media. Quite a scandal! The founder-CEO lost his job, and the company immediately vanished from our competitor list.

BOOKKEEPING AND TAXES

General day-to-day bookkeeping is both easy to do yourself and easy to outsource. If you do hire an accountant, note that, as with law firms, smaller accounting firms charge significantly less than large firms.[4] Whether you engage a professional accountant or not, you will have to procure one of the many affordable accounting software packages, set up an account structure, and start making journal entries. Accounting programs force the user into double-entry bookkeeping and other best practices. Most of them were invented for a good reason, but you don't have to understand them in detail to apply them. The software will do that for you.

Filing personal tax returns is a nuisance, and filing corporate tax returns as a young entrepreneur isn't any better. Fortunately, you won't be paying much in corporate income taxes right out of the gate. So just make sure to file on time and stay out of trouble.

If you are operating as a C corporation and you are in the happy situation that you are cash flow positive, watch out for double-taxation traps. You avoid them by paying all of the cash left over at the end of the year to executives and employees as bonuses reported on W-2 forms. That way the company does not report any profits and hence doesn't owe any corporate taxes.*

If you are operating as an LLC or S corporation, the IRS offers an incredible tool to keep taxes for your small business at a minimal level. You have the option between *cash*-based and *accrual*-based tax accounting.

Cash-based accounting means that you pay taxes on positive operating cash flow between January 1 and December 31 of a given calendar year. It doesn't really matter how the cash flow was generated.

---

* There are certain laws on excess compensation that can get in the way of this approach, but they typically do not apply for early-stage bootstrapping startups in survival mode.

It does not matter whether you are behind in deliveries, whether your customers owe you long overdue payments, or whether you got a huge advance. All that matters is how much cash flowed in and out of the business. Hence, if you are trying to save on taxes in a given year, all you need to do is reduce your net cash flow. Surprise your customers by asking them to delay paying their bill into the new year! Surprise your vendors by paying for services early!

Accrual-based accounting, on the other hand, is based on the timing of the exchange of services and products, not the timing of payments. If you sell a product in December and the customer has not paid by December 31, the income associated with the sale is still taxable in the tax year the product was shipped. If you bought and received a service in December but you have not paid for it by December 31, the expense is nevertheless deductible in the year the service was rendered to you.

The IRS restricts the use of cash-based accounting. If your company is making too much or if your business carries inventory for sale, you are not allowed to opt for cash-based accounting. But assuming you can avoid those two restrictions, cash-based tax accounting can be extremely beneficial to small cash-flow-positive enterprises.

In addition to filing for and paying income taxes, you are also responsible for collecting sales taxes on behalf of your state. And as if that weren't enough, the moment you pay anyone a salary, including yourself, you will have to withhold income taxes for the employee, which is what the next section is all about.

PAYROLL AND MANDATORY BENEFITS

It is easy to get overwhelmed by the complexity of running payroll, withholding taxes, and providing mandatory employment benefits. I certainly was anxious when I needed to issue ThingMagic's first paycheck (to myself no less). As it turns out, these regulatory requirements can be squared away by using one of many payroll service providers. Payroll companies offer their services to even the smallest of firms for reasonable fees. Trying to process payroll in-house instead would be nothing short of suicidal for a small venture.

Payroll companies offer any or all of the following services, typically in a bundle:

- Weekly, biweekly, semimonthly, or monthly payroll processing (Note that some states have laws on the minimum frequency of payroll processing.)
- State and federal income tax withholding
- Paychecks or direct deposits to the employees' bank accounts
- Bonus payments and other extraordinary payments
- Reimbursement checks for travel and other expenses
- Deductions for retirement accounts
- Deduction for flexible spending accounts (FSAs)
- Issuance of W-2s at year-end
- Other pre-tax deductions including the health insurance employee contributions
- Increased payroll withholding (per employee request)
- Final payroll payment upon employee termination

To differentiate themselves, some of the payroll providers mix additional services into the package. The sales pitch for the extended offering typically suggests that the package will cover your HR function in its entirety. However, elimination of the internal HR function is unrealistic for many reasons, while the added services may cost you dearly. Here are some of the supplementary services you can buy:

- HR hotline
- HR-related legal services
- Multistate payroll processing

- Labor and benefits compliance and management

- Policy and handbook support

- Performance management

- Compensation management

- Recruiting and selection

- Risk and safety management

- Training

- Leadership development

- Retention management

INSURANCE

You can buy virtually any type and any amount of insurance coverage for your fledgling company, and yet there is no coverage that would keep your vulnerable little business out of ruin if disaster were to strike. Given that limitation, put all the mandatory insurance coverage in place, but take it slowly as far as optional coverage is concerned. There will still be time to buy million-dollar policies, when you are worth millions of dollars.

Getting quotes from insurance companies and brokers is free. The more requests for quotes you send out, the more likely you will get the best deal in town. A good broker can help you assess the risks inherent in your business, sort through the various options, and ultimately save you a ton of time:

- **Workers' compensation insurance** benefits are mandatory and fully regulated by the government. You should get the same deal irrespective of the insurance company you pick. Workers' comp pays benefits if an employee gets hurt at work. The premium is determined by the mix of employees, roles and job descriptions in the company, and by the type of business. Fortunately, high-tech companies tend to be on the low end of the premium spectrum.

- **General liability and property insurance** will be required the moment you get yourself a landlord or engage with a significant client. Without property insurance, you won't get a lease, and without liability insurance, no client will give you a substantial contract. Usually, you can get basic coverage and then deal with additional coverage requirements using insurance riders.

- **Directors and officers insurance** is usually required in venture-backed companies. The premiums are significant and annoying, but there is a positive implication for the founder group: since the insurance is offered as a package for the entire company, founders, in their capacity as executives or directors, benefit from the same protection.

- **Key person insurance** essentially works like life insurance except that the company pays the premiums, and it is the named beneficiary. Companies buy key person insurance to mitigate the risk when a founder or key employee dies or gets incapacitated. Money can't really replace a founder, but it can help offset a difficult situation. Premiums depend heavily on the age and health of the insured.

- **Errors and omissions insurance** is a specific form of liability insurance insuring the actual performance of a service or the functionality of a product. Policies are very expensive to buy, and young companies usually pass on this one . . . in part because no insurance company would consider writing such a policy for a startup without a track record.

- **Travel insurance** covers employees on business trips, mostly ensuring that they will be brought back home should they get sick or die.

- **IP infringement insurance** protects a company in the case of litigation related to intellectual property. It's not cheap, but it may make sense in certain litigious industries and situations.

- **Employment practices liability insurance** protects a company and its officers against claims from employees, former employees, and potential employees. The claims covered include employment-related allegations including discrimination, sexual harassment, and wrongful termination.

  Alternatively, you can protect yourself against employee claims by treating your staff well. Why not spend the money you save in insurance premiums on a company outing and a few well-deserved spot bonuses?

OFFICE SPACE

Your young technology company, however small it may be, needs a home. In fact, the smaller the company, the bigger the need to muster the collective energy of the team and create an atmosphere where people inspire and help each other through problems, obstacles, and moments of doubt.

Garages

Working out of a garage, home office, or other private property is a great way to get your technology company started. However, be aware of some common pitfalls for a young company occupying a cofounder's house:

- **You should cohabitate, at least during normal office hours.** Young companies need the energy generated by a group of people creating something new together. At a time when the future of the venture is unclear and cofounders are sorting out their personal career and financial future, physical proximity is an important stimulus. Hence, try to consolidate your team as much as possible in one office, as unconventional as that space may be. If you choose a garage, make sure that everybody is comfortable enough to show up there for work.

- **Don't abuse your cofounder's hospitality.** If a team member is generous enough to offer her house as a temporary office space, make sure to not overstay your welcome. Don't abuse the generosity to the point where your company becomes a burden on the landlord and her family, lest you end up losing a valuable colleague and your company's living quarters all at once.

- **Don't be a slumlord.** Employees can work under constrained circumstances for a while, especially in pursuit of a worthwhile and challenging goal. However, at some point the uncomfortable office environment will become a limiting factor. As a founder or team leader, you need to observe when your team's productivity starts to suffer from the office space situation and when the lack of more space becomes an obstacle to your company's success and growth.

If you run into any one of these issues, go find yourself a bigger garage. Make sure the new space includes a bathroom, insulation, and a working heating system. Access to the new office should not require climbing over your cofounder's couch. There should be enough chairs and desks so that nobody has to sit on somebody else's lap. Open space is fine and good for the team spirit, but make sure to provide for some reasonably sound-insulated corners so that your employees can have a private phone call or meeting.

Shared Facilities

In many ways, a full-service shared office facility is the complete opposite of the garage approach. Think living in a luxury condo versus a tent. Yet, when you decide that you'd rather travel to a dedicated office space each day as opposed to working at your dining table, it is the best option for your growing high-tech venture. At ThingMagic, we leased a 400-square-foot space at the Cambridge Innovation Center (CIC),[5] a full-service facility for young companies, about a block from the MIT campus.

CIC was started as an incubator, but it had to dramatically adjust its business model after the dot-com bust. Since then, the company has been acting as a landlord and service facility for startups and other ventures in need of smallish facilities. CIC has quickly become an important element of the Cantabrigian and Bostonian startup scene and a role model for many similar centers around the world. The model is based on a few important insights:

- **Maximize flexibility.** Startups need flexibility when it comes to quickly increasing or reducing the space they are occupying. Hence, CIC established flexible spaces, divided by sliding walls. Four to five people fit into a space. When a startup needs to expand, the next bay is opened up. Since the spaces are configured in similar ways and contain the same furniture, a move or extension is easy for both landlord and tenant.

- **A good location is worth more than a few extra square feet.** CIC is located in the business district of Cambridge, next to MIT's campus, close to lots of other high-tech and biotech businesses, and has direct access to one of Boston's main subway lines. The location helps young companies find employees who are looking for an easy commute or need to interface with collaborators at MIT.

- **Why own your own copier if you can share?** Resources such as conference rooms, printers, copiers, IT infrastructure, networks, food, and telephones can be shared. The preferred conference room may not always be available when you want it, but let's face it, that would not be any different in a private office.

  The big advantage of shared resources is that a company gets access, no matter how large it is or how much space it rents. In fact, some companies maintain shared space for the sole purpose of being able to hold customer meetings.

- **No need to be lonely.** Early-stage entrepreneurs can get lonely. They don't usually have a big team of employees to converse with, and their families may have already written them off as lunatics. Chatting with fellow entrepreneurs and outcasts in the shared kitchen helps maintain some degree of social interaction.

Private space in a shared facility easily costs twice as much per square foot as regular office space. However, as you consider the two options, remember to factor in the cost of all the benefits that the shared facility provides without charging extra, including the actual management of the space. When you compare apples to apples, you find that shared facilities can be cost-competitive even for larger organizations with more than a hundred employees.

Subleases

In 2001, lots of former dot-com spaces became available in and around Cambridge. We stood in awe, visiting the various sublease options, built in the most spectacular fashion, many only months before our visit. The overwhelming extravagance included gigantic pool tables, bar areas with velvet sofas, double-height round staircases, and sound-proofed boardrooms that would have made the CIA proud. Fortunately, these rather useless amenities were surrounded by fully equipped cubical and office areas, and many of the spaces could be had at discounts of up to 80 percent off previous market rates.

When demand for office space drops, large landlords do not release their tenants from long-term leases. A large company that all of a sudden has tens of thousands of square feet of unused space is forced to pay rent until the end of the lease, which could be anywhere from one to nine years. However, the company has the right to sublease, and it will typically take any deal that helps it cover some of the financial burden.

Sublease agreements tend to be much more creative and accommodating than regular leases. Before you agree to terms, especially in a down market, insist on special provisions including the following:

- Free rent for the first 12 months (depending on the length of the lease)

- Increasing rent over the term of the lease (starting with a very low rent in the first year)

- Furnished space including cubicles, desks, chairs, and IT equipment

- Included utilities (electricity, cleaning, heat, and so on)*

None of us wants to deal with another economic crisis anytime soon. However, should bad economic times strike again, at least you can expect to find interesting sublease opportunities.

Assuming you do not inherit furniture from the last tenant or the landlord, regular furniture and cubes can cost you dearly, unless you get creative: a wooden door and four desk legs from Ikea can be had for about $60. Once you assemble these materials and add a desk chair from a used-furniture store, you have all the furniture you need.

When ThingMagic moved into its first regular office space, a few of my colleagues literally cleaned out the doors of every single Home Depot store within 20 miles of Boston. We also invented 10 different ways of cutting and finishing the doors to create different types of desks that would enable a perfectly functional and pleasant work space!

---

* At the time of this writing, the commercial real estate market in Cambridge is having an unprecedented boom. Inventory rates are at historic lows, and rents are skyrocketing. Don't count on finding any bargains here right now!

HOUSEKEEPING

Lesson 1: Hire a smallish law and accounting firm. Small firms charge significantly less, while providing more personalized service.

Lesson 2: Evaluate carefully whether to incorporate as an LLC, an S corp, or a C corp, and discuss your plans with a lawyer.

Lesson 3: Never let compliance issues stop you from building the company you want to create. Just quietly do what the law requires you to do, be diligent about it, and recognize that your competitors have to go through the same tedious process.

Lesson 4: Engage a payroll processing firm to run your payroll.

Lesson 5: Buy only mandatory insurance coverage for your young startup because no insurance will protect you from the real risks threatening the livelihood of your company.

Lesson 6: Shared office spaces with communal resources are excellent choices for young technology startups. Not only are those spaces more economical overall, they also relieve the founding team of many administrative tasks and worries.

CHAPTER 7

Compensation

Money can't buy happiness, but neither can poverty.
—LEO ROSTEN (1908–1997)

A young venture relies on the commitment of the individuals to the company as much as it relies on the commitment of the company to the individuals. There will come hard times for the company, and there will come hard times for the individuals driving it. If the startup and a key employee get into trouble at the same time, it can mean the end of the venture. Short of such an unfortunate meltdown, the resources of the company can help stabilize the individual. Likewise, a well-balanced and grounded founder or executive can make a huge difference in managing the young company through crisis.

As you come up with a compensation structure, remember to take good care of yourself and your early employees, so that they will take care of you and the company.

EQUITY: A STARTUP'S MOST IMPORTANT CURRENCY, BUT WHAT IS THE EXCHANGE RATE?

When you incorporate your venture, you and your cofounders own a large majority if not all of the company. It's a great starting point, and yet, paradoxically, such a clean capitalization table helps you get off the ground only if you are willing to give away some of the precious equity.

You can ask your first employees to work for below-market salaries only if you incentivize them with stock or option awards at the same time. You can attract angel investments only if you are willing to part with preferred stock or issue convertible notes. Rather than holding on to all the stock, building a startup requires giving away equity. Your goal should be to give it away at a slower pace than the stock is increasing in value.

For some startup employees, equity awards are considered compensation in lieu of salary. For others, equity is the only reason to join a startup. Equity is what you don't get when you work for a large corporation. Equity is what makes one rich. Equity can be life changing. Before you start making everybody happy with your generosity, let's look at the different kinds of equity-based instruments there are to give away.

Restricted Stock

Restricted stock refers to stock that is subject to certain restrictions, such as a vesting schedule or limited transferability, and that is subject to risk of forfeiture when the recipient leaves the company before the stock is vested. Restricted stock is typically given to founders and key team members at a time when the company is worth very little. The stock must then be "earned" over a period of several years. Restricted stock awards offer significant tax advantages compared to other equity instruments.

Immediately after incorporation, the company decides how much stock it is going to issue. At the time the stock can be issued at the par value, which may be as little as a fraction of a cent.* While the recipient of such founder stock is required to pay for the grant with

---

* This assumes that the fair market value of the company is effectively zero at the time of the company's founding. The fair market value may be higher if assets, such as intellectual property, are transferred to the company.

post-tax money, the total amount due is negligible even for a founder coming right out of school.

Within 30 days of issuance and purchase of the stock, the employee files an 83(b) election with the Internal Revenue Service. By means of the 83(b) election, the recipient of the stock declares that he pays income taxes on the value of the stock at the time of the award rather than upon vesting of the stock.* Years later, when the stock is sold, the individual pays taxes on the gain—that is, the difference between the fair market value at the time of award and the sales price.

Importantly the 83(b) election starts the capital gain's clock on the entire stock grant. After one year and one day from purchase, the stock qualifies as a long-term capital investment. If and when the stock is sold after that date, the grantee will be subject to the more favorable capital gains tax rate, as opposed to ordinary income tax.

Restricted stock in early-stage companies is usually subject to reverse vesting provisions. Even though all of the stock is awarded on day one, the company has the right to buy back the unvested portion of shares at the original grant price if the grantee departs before the end of the vesting period.

In later-stage companies, issuance of restricted stock becomes more difficult. Once the company has gone through a financing event, every share of stock will then have a well-defined value. At that point, a restricted stock award can mean a significant financial burden for the recipient. Most employees and executives are not excited about paying money in exchange for a security that may not be worth anything in the future. Nevertheless, it is possible to issue restricted stock to employees in a later-stage company:

- **Option A: We generously allow you to buy our stock.** The company sells stock to the individual at market value. The employee is required to pay the fair market value of the stock in cash at the time of purchase. However, the employee's purchase puts that cash at risk. If the company goes under or if the stock price at

---

* To be precise, the recipient pays income tax on the difference between the price and the fair market value of the stock at the time of award.

exit is lower than it was at the time of the award, the recipient will lose the invested money.

- **Option B: We give you the stock for free, but we can't help you with the taxes.** The company sells restricted stock to the individual below the stock's fair market value, for example, at par value, which may be nominally zero. In this case, the recipient gets away with paying the company only a few dollars. However, the IRS recognizes that the recipient received compensation in the amount of the fair market value of the stock at the time of the award. Therefore, the individual has to pay income tax on the fair market value of the stock (less the amount actually paid to the company).

In both cases, the recipient of restricted stock benefits from long-term capital gains taxes on gains if the stock is sold more than one year later at a value higher than the fair market value at the time of purchase.

Some companies come up with complicated loan structures to allow senior executives to buy restricted stock when that stock has become expensive. For example, the company may decide to grant the individual a loan to buy restricted stock. Such compensation was popular when it looked like the value of a high-tech startup could only go up.

Unfortunately, these structures can result in disastrous situations when things don't go as planned. If the stock loses in value after the award, the executive then owes the company a lot of money, and the company, acting in the best interest of the stockholders, has to call in the loan. While it is tempting to create complex instruments so that the employees can benefit from the preferential tax treatment of restricted stock awards, the risks for the individuals are considerable.

Incentive Stock Options

As the name suggests, *incentive stock options* (ISOs) are instruments to incentivize founders, executives, and employees of the company. The

IRS has established special tax rules for ISOs that significantly reduce the tax risk for the ISO recipient compared to other equity instruments. ISOs can be awarded to and held by only current employees of the company.

Options are issued at a specific strike price. The strike price of an ISO needs to match the fair market value of the underlying stock at the time of purchase. However, ISOs are typically options to purchase common stock, and common stock can be priced much below the preferred stock an investor would buy.

The argument for a low company valuation for the purpose of keeping the option strike price as low as possible is one of the few occasions when you want to go all out trashing your company. Go ahead and express the frustration that has built up since your company's founding: little revenue, weak market, tough competition, and delayed product releases! Anything negative you can think of, you bring it up when you are justifying the pricing of your stock and stock options.

The second argument you need to make relates to the difference in value between preferred and common shares. Ironically, the bigger the gap, the more attractive your ISOs! Preferred stockholders are usually entitled to a long list of special rights including special voting rights, redemption rights, board seats, conversion rights, anti-dilution provisions, rights of first refusal on transfers, information rights, and—most important—liquidation preferences. These provisions make preferred shares significantly more valuable than common shares. Common stock can be valued at a fraction of the preferred stock, which can be especially important when your company is going through rough times and you don't see how you might ever manage its valuation above the liquidation preferences.* The lower you can push down the common stock price and the strike price of the options, the more attractive the ISOs will be for your employees!

Every time stock options are approved by the board, you need to document a recent valuation exercise—that is, a 409A valuation. You

---

* What is considered a reasonable difference in fair market value between preferred and common shares is changing constantly as a function of economic conditions and guidelines provided by the IRS. Do consult with an accountant or auditor to discuss your specific situation.

should revalue the stock at least once every 12 months—more often if there is a significant change in the business in the meantime.

ISO holders exercise their options in three possible scenarios:

- **"Working for a successful startup is just great!"** If the company incurs a change of control event (liquidity event) during the ISO owner's employment, he is paid the difference between the market value of the underlying common stock and the strike price of the vested ISO. The owner pays ordinary income taxes on the gain if the consideration is cash. If the consideration is stock that is held for a year or more, the owner may get the benefit of the capital gains tax rate. If the value of the stock is below the strike price, no money is paid out, and no taxes are due.

- **"I'm leaving, but I trust they'll be successful without me."** If the ISO holder's employment with the company ends, the employee has three months from termination to exercise his options.* Before the end of the exercise period, the departing employee needs to make a judgment call on whether or not the stock will be worth something in the future. In some cases the decision is easy, but more often than not, the departing employee shies away from paying serious money for stock, the real value of which is utterly unknown.

- **"It's risky, but I'd better make some moves to save on taxes."** The holder of vested stock options may decide to exercise his ISOs in order to start the capital gains clock on the stock. At the time of exercise, the difference between strike price and fair market value is subject to the alternative minimum tax (AMT), but not to income taxes. When the employee sells the stock in the future, he owes capital gains tax on the difference between strike price and sales price, as long as at least one year has passed since exercising the ISO. If less than a year has passed, the gain would be taxed as ordinary income.

---

* The company and the parting employee can agree on an extension of the exercise period beyond three months; however, in that case the ISOs automatically convert to NQSOs (discussed in the next section).

Unfortunately, the strategy can backfire badly if the stock drops in value between the exercise and sales dates. The dot-com bust forced certain startup ISO holders into bankruptcy when they were required to pay AMT tax on stock that was no longer worth anything.

Nonqualified Stock Options

Nonqualified stock options (NQSOs or NSOs) are options without the restrictions and benefits of ISOs. They can be issued to any individual for services rendered, including consultants and advisors of the company. Like ISOs, NQSOs typically come with a strike price that equals the fair market value of the underlying stock at the time of the award. In most scenarios the NQSOs are exercised on the occasion of a liquidity event. The owner receives the difference between the stock price and the exercise price and pays ordinary income tax on the received money. NQSOs carry an expiration date, but they do not have to be exercised within three months of the date that the recipient's employee relationship with the company ends.

Nonqualified stock options are useful instruments to compensate a founder or other affiliate for a particular, critical, or risky service. For example, if the personal financial situation of one founder allows him to grant a significant loan to the company, that willingness to risk his own money for the benefit of the company deserves the issuance of NQSOs. Or if a founder makes significant tangible or intangible property available to the company, for example, significant IP, such generosity is best rewarded with an NQSO grant.

By keeping the compensation for such contributions separate from founder stock and salaries, the company can be quicker and more flexible in its response to extraordinary contributions, while avoiding a class system among founders.

Warrants

Warrants work like NQSOs, but they are used in a different context. Like NQSOs, they typically are exercised at the time of a liquidity event, and they are subject to regular income tax.

Warrants are mostly used to create additional upside for investors, creditors, or channel partners. If a bank grants a loan to the risky startup, warrants are issued to the bank to sweeten the deal. Warrants are also used to motivate liquid investors to invest more than their pro rata share in a follow-up funding event, even though the company's performance might be lagging.

Warrants are not governed under the company's stock option plan and therefore offer more contractual flexibility. For example, warrants can be issued on common or preferred stock.

VESTING, CLIFF VESTING, AND ACCELERATED VESTING

Equity grants are subject to provisions that protect the grantor in such a way as to hopefully get maximal value from the grantee in return. Most notably, equity grants are subject to vesting provisions, which means the individual earns the equity over time and in exchange for services rendered.

Vesting: No Free Lunch for Anybody, Not Even the Founders

While there are few regulations or specific rules regarding permissible vesting schedules, it is well worth thinking the issue through carefully. The standard vesting schedule is laid out in the equity incentive plan and can be varied on an individual basis in the grant agreements. Different plans can be created for executives and rank-and-file employees, but it is a good idea to keep vesting schedules as consistent as possible.

The vesting period for founder stock should be based on two considerations: How long do the founders intend to run the company? How long will it realistically take to make the business successful? If the period is too short, the team will have no time to build something valuable, and the founders will reach the end of the vesting period without motivation to keep going. Staying with the company won't hold much, if any, additional advantage. On the other extreme, if the vesting period chosen is too long, the founders may not feel that being there and working hard are sufficiently rewarded. A founder might leave because he is not sufficiently vested.

The former problem is more serious and harder to fix after the fact. An overly long vesting period, on the other hand, can be shortened easily with a provision for accelerated vesting, as we will see below.

Vesting periods of more than four years are unusual. However, there is really no good reason why vesting shouldn't extend to five or six years. Founders and employees should realize the benefit of getting stock awards early on and accept the notion that they will have to stay longer to earn it all.

Vesting typically happens in monthly or quarterly increments. At the end of each vesting period, the holder of ISO options has the right to exercise the additional options, while the holder of restricted stock has the right to keep all of his vested stock. Monthly increments avoid artificial nonlinearities that may cause the employee to *play* the system. If the employee is unhappy, he might as well leave rather than wait for the end of the quarter to get the award. Longer vesting intervals, on the other hand, lower the administrative burden for the company.

Cliff Vesting: No Rewards for Quitters

Option and stock plans typically provide for a *cliff* on the first or second anniversary of the employee's start of employment. No vesting occurs until those dates.

Cliff vesting helps to prevent the scenario in which a new employee becomes a stock owner but stays for only a very short amount of time. For both administrative and fairness reasons, companies do not like to award stock to hires who didn't work out. Instead, the company prefers to sever the relationship completely and not pay *alimony*.

Cliff vesting enables a trial period for both the company and the employee. A cliff vesting clause makes sense for founding teams as well, especially when there is uncertainty about who will stay with the company after the initial company formation. For founders, the initial vesting can take into account services provided prior to the actual vesting start. If founders contribute greatly before the official start of the company, some portion of the founder stock should be declared vested at the time of incorporation.

Accelerated Vesting: Let's Be Generous, if Things Work Out

Accelerated vesting provisions typically kick in at the time of a change of control of the venture. There are two main reasons to include accelerated vesting provisions in equity awards:

- **To avoid injustice at the time of a liquidity event.** For example, consider a member of the management team who joined very recently and therefore owns no vested stock. Yet, the executive was instrumental in lining up and executing the acquisition of the company. The accelerated vesting provision ensures that a portion of the executive's stock becomes vested as part of the liquidity event.

- **To motivate key people to negotiate aggressively for a favorable exit.** A situation in which key employees have second thoughts about the timing of an exit simply because they are not sufficiently vested could be very detrimental to the outcome of an M&A discussion.

Accelerated vesting provisions make a lot of sense when used properly and in moderation. However, if used selectively for the benefit of certain stakeholders, accelerated vesting may lead to significant injustice among team members. A specific clause can be easily hidden in the legal verbiage. Consider the following case:

The two cofounders and the CEO in the hypothetical JonCo, Inc., receive the respective equity stakes of 40 percent, 40 percent, and 20 percent. The shares are subject to linear monthly vesting over four years. In addition, the CEO has negotiated a 100 percent accel-erated vesting clause for himself, claiming that that is the industry standard for executives of his caliber. The company is sold after 12 months. Due to the accelerated vesting provision, the effective owner-ship at the time of the change of control would be 25 percent for each of the founders and 50 percent for the CEO. This certainly wasn't the intent of the initial equity award to the founders.

The founding team should keep in mind that executives negoti-ating for a package who make claims about what is typical may very well be overstating their case. What is typically done may or may not make sense in the particular situation.

There are an infinite number of variations of accelerated vesting provisions, but typically the provision looks like one of the following statements:

- Upon a change of control, 50 percent of the unvested shares shall immediately vest.

- Upon a change of control, vesting shall be accelerated by 12 months.

The definition of "change of control" typically includes the sale of the company or its assets, a merger, a change in the board compo-sition or election rights, or the liquidation of the company. An IPO usually does not trigger accelerated vesting, mostly because the under-lying stock continues to exist and employees simply continue to vest after the IPO. In fact, the company needs its employees after the IPO more than ever, and there is no better retention tool than unvested pre-IPO stock and options.

The equity incentive plan usually does not provide for any auto-matic acceleration in vesting. Rather, acceleration rights tend to be negotiated on an individual basis. At the time of a change of control, the plan will generally allow the board of directors wide discretion to accelerate any or all options that remain unvested, provided the buyer agrees.

CASH AND BENEFITS

Cash is hard to come by in a small high-tech company. It is difficult to earn and expensive to raise. That's why young ventures attract talent by offering equity rather than lavish salaries whenever possible. And yet, most of us need cash on a daily basis to eat and live a dignified life. For practical purposes, startups need to compensate their employees with sufficient salaries and benefits so that the employees don't have to seek employment somewhere else. Call it twenty-first-century high-tech Taylorism!

Cash Compensation in Cash-Strapped Ventures

Salary expectations and needs among the founders and the early team can vary greatly. The young university graduate will find almost any salary outstandingly high compared to the student stipend he received until very recently. The senior executive with spouse, kids, and mortgage, on the other hand, will be much harder to please.

Early on at ThingMagic, we established the principle that every founder and key executive received the same salary compensation and participated in the deferred compensation program when necessary. This meant that the CEO would be undercompensated in terms of cash and that others would get relatively generous pay packages. We then used equity to adjust for the different levels of responsibility, experience, and start of employment.

This approach of equal cash compensation and equal *suffering* proved helpful for the company's cash flow and the morale among the management team. Which executive would dare to complain about his pay package when the CEO was making the exact same amount? It also became a lot easier to argue against salary increases. What may seem like a small amount of extra money when given to one individual can become a significant burden for the company when the entire team gets the increase.

A startup's cash outlook early on is uncertain at best. Founders and the early executive team alike need to understand that there is no way to predict how much money will be available for compensation, especially pre-equity funding. Rather than risking bankruptcy caused by large contractual salary payments, modest salary arrangements with a larger bonus are the safer and more constructive way to go. There is no rule that a bonus can't be paid on a monthly basis, nor are there rules or restrictions on what metric can trigger the bonus payments. As long as there is a clear and unambiguous understanding about how a bonus is calculated, most can live with a creative arrangement.*

Some contributors, founders, or executives are in the fortunate situation to not need cash compensation at all. These individuals are rare and hard to come by though. What's more, having a lot of existing personal net worth doesn't exactly motivate someone to work hard. Hence, when someone offers to work without cash compensation, take him on. However, be prepared for the executive to be worrying more about his boat, oceanfront house, and Porsche than about the daily problems and future of your company.†

Benefits

Conventional wisdom suggests that small companies offer significantly fewer benefits than large corporations. With a little bit of creativity, this doesn't have to be the case. While certain key benefits are expensive, many types of benefits provide value to the employee without financially burdening the small company. Some benefits are even free for all practical purposes.

Even if your company does not have the money to contribute much toward benefits, you can do your employees a great service by handling a good part of the thinking for them. Young employees,

---

* Make sure you are not violating minimum wage laws!

† Watch out for wage and hour laws that may carry triple damages if violated. Of course, minimum wage requirements are not your first thought when you hire a millionaire who cares about equity only. However, you never know where the relationship is ultimately going. He who seems a generous and friendly contributor today may just end up battling you in court when the company or the relationship deteriorates.

notably engineers, tend to be focused on their work and oblivious to the practical matters in life, including such critical issues as health and health insurance. This is good news because you want your employees to work rather than spend valuable time browsing for life insurance providers! If you can offer them easy-to-enable options for critical benefits, you are helping the employees and the company, regardless of who pays the premiums!

The following benefits are either critical for the individual or inexpensive for the company, or they have a big impact on the quality of life for the startup's young employees.

Health Insurance. This benefit costs dearly no matter how you twist and turn it. Under a typical corporate healthcare plan, the employer pays a majority percentage of the premium, and the employee pays the rest. However, this model is more or less arbitrary, and other plans with less or no employer contribution are equally valid. Importantly, both the employer and the employee contributions are pre-tax contributions. Hence, no matter how the premiums are split, the net cash outlay between the employee and employer is the same. Of course, in a competitive market, job candidates will look at the overall compensation package including the coverage of healthcare premiums very closely.

The Affordable Care Act (ACA) has made it easier for small enterprises to get reasonably priced coverage for its employees. Entrepreneurs are now able to shop for insurance on online portals. The reform is supposed to ensure the portability of insurance and the ability to switch from plan to plan. It is also supposed to offer protections in the case of preexisting conditions.

Many of your employees will have the opportunity to get health insurance coverage through their spouses or partners. Since this can save you serious money, you may want to motivate your employees to take that approach. At ThingMagic we used to pay employees an extra $100 per month if they declined our health insurance and got it through someone else's plan.

Vacation Time. U.S. employers are notorious for offering little vacation time compared to much of the rest of the world. Startups expect employees to work nights, weekends, and whenever there is a need. It is therefore only fair that those very same employees get sufficient vacation time to recover from hard work and unpredictable work hours.

As important as the number of vacation days is the carryover policy from one calendar year to the next. If you don't allow carryover, employees are forced into using their vacation time before the turn of the year, possibly at a very inconvenient time for the company. If you do allow carryover, employees tend to accumulate vacation time, don't get their much-needed rest in a given year, and build up a vacation liability on the company's balance sheet.* A good compromise of offering a limited number of carryover days goes a long way toward addressing both problems.†

Workers' Comp Insurance and Unemployment Insurance. These are mandatory benefits in the United States. Fortunately, the monthly premiums are modest.

Retirement Benefits. Retirement savings and benefits programs come in many different forms. In the United States, the most common employer-sponsored program is the *401(k) program.* Unfortunately, fees and administrative overhead expenses for 401(k)s are significant. Small companies offer *simple individual retirement account* (IRA) *programs* instead. Simple IRAs are very similar in structure to 401(k)s, but they are cheaper to set up. In either program, the employee makes monthly contributions and gains accrue on a tax-deferred basis. The employee will be able to withdraw money penalty-free after age 59½.

---

* Unused vacation time ends up on the balance sheet of the company as a liability. Most startups are not really concerned about that since cash flow is not affected . . . until the employee leaves the company, which is when the company has to pay for the outstanding vacation time all at once. That can hurt quite a bit!

† Some companies have adopted the policy of unlimited vacation time. The idea is that employees take vacation as they deem appropriate but not abuse the policy. The policy does avoid the carryover problem, but one can't help to suspect that the policy causes employees to not take sufficient time off for fear of losing their jobs if they did.

The employee's paycheck contribution can be matched by an employer contribution, but matching is not mandatory. More important than matching is the fact that the company-sponsored program enables the employee to save toward his retirement and to take advantage of the preferred tax treatment.

Another retirement option for small businesses is *simplified employee pension* (SEP) *IRA accounts*. Contributions to SEP accounts are made solely by the employer. The contribution limits and qualification requirements are different from simple IRA accounts, but the benefits are very much the same: tax-deferred accrual until the money is withdrawn after age 59½.

Disability Insurance. This insurance is mandatory in some U.S. states, including California, Hawaii, New Jersey, New York, Puerto Rico, and Rhode Island.[1] Everywhere else it is up to you to offer a program or not. Bad luck can strike anyone at any time, even the young professional. The earlier in your career you get disabled, the more disastrous are the consequences if you and your family are not protected.

Life Insurance. This is not the first thing on employees' minds when they start their careers in a small company and are surrounded by a bunch of other young folks. Basic life insurance benefits, however, are not all that expensive, especially for a young employee pool. As with other types of insurance offered, there is no need for the company to pay the premium. It can be left to the employee to make the decision as long as the company offers the program.

Flexible Spending Accounts (FSAs). These accounts cover healthcare costs and dependent care costs. Apart from a nominal program cost, there is no cost to the employer, but there are significant tax benefits to the employee.*

---

* Technically, the company fronts the annual FSA account total, while the employee gets a monthly deduction from his paycheck. Since the employee has the right to use the money all at once and then quit, it is possible for the company to spend a significant amount of money on behalf of an individual who leaves his job early in the year.

Health Savings Accounts (HSAs). These are similar to healthcare FSAs in some aspects, but they differ in others. Importantly, they are not available with all health insurance plans.

Commuting Expenses. Companies can elect to pay for the employee's commuter train or bus ticket. Not many startups are able to run their own bus system, as a few Silicon Valley giants do, but most startups should be able to afford their employees' monthly subway passes.

Bananas and Other Healthy Food. The sushi lunches at Google are legendary. Unfortunately, these perks provided by highly successful technology companies make it sound like only rich companies can afford to invite employees to eat in. In reality and by comparison with other perks, food is not expensive, but it is fully tax deductible as long as the company feeds its employees for the convenience and benefit of the business. Paying for a busy evening or weekend in the office with a meal costs a lot less than paying overtime.

In summary, careful planning of benefits can help save a lot of money on the part of the company and the employee. In a small company, paying for benefits is mostly a zero-sum game. It ultimately doesn't matter if the employee or the company pays because the money comes from the same pot. However, the beneficial tax treatment of employer-sponsored benefits can have an enormous impact on the overall cost of the package.

In a small startup, most everybody's needs are pretty well known. Tailor the benefits package to the specific needs of founders, executives, and employees, and adjust the offerings when opportune. Don't put out food if everybody is always traveling, and make sure to add life insurance when you start sending employees to war zones.

DEFERRED COMPENSATION

Entrepreneurs happily sacrifice salary and perks that could be earned today for a much bigger payday at an unknown later time. On a shorter time scale, *deferred compensation* refers to delayed cash payments, salary payments, bonuses, and consulting fees. Deferred compensation is money at risk for the employee, but if used smartly, it can help reduce the impact of a cash crunch and mitigate risk for the young venture. Deferred compensation reduces the company's exposure to market uncertainty, its dependency on investors, and its equity dilution due to third-party investments. There are multiple mechanisms to defer cash compensation for founders, executives, and key employees: *

- **Deferred W-2 compensation.** Employment law is complicated, and it is designed to protect the employee, including and in particular with respect to wage and salary payments. Agreed-upon salaries are to be paid and are to be paid on time. Bonus payments, on the other hand, even though reported on W-2 forms, come with quite some regulatory freedom. If you need flexibility as far as payment schedules are concerned, make sure to partially compensate key employees through bonus payments. You can set the rules and trigger points for bonuses as you wish, but be careful to establish a structure that you can actually follow through with and that complies with wage and hour laws.

- **Deferred partnership income.** Members in partnerships such as LLCs receive partnership income in the form of guaranteed payments, which is taxable for the individual as self-employment income. Guaranteed payments can be made at any time to any member of the partnership. By the same mechanism, distributions can be delayed, deferred, or canceled, subject to the IRS's deferred compensation rules and wage and hour laws.

- **Founders as creditors.** Formal notes of credit issued to founders and key employees can be used to get around cash flow timing issues. Founder loans help navigate short-term tax and regulatory

---

* Deferring compensation has gotten more complicated since the addition of Section 409A to the Internal Revenue Code (Section 885 of the American Jobs Creation Act of 2004). Definitely consult with an accountant or lawyer to make sure you understand the detailed regulations.

requirements at the end of the fiscal year. They can also be used in lieu of bank credits, which are usually not available to startups or are extremely expensive.

When founders or employees are lending money to the company, the loan should be documented with the same formality and diligence as a bank loan. For compliance reasons, the company needs to be able to substantiate the validity, legality, and timing of a loan. This means—at a minimum—that the note states the principal, maturity date, and an interest rate consistent with market rates. Also, the founder would want to be assured that the money owed to him will be paid back eventually. Circumstances at the company can always change, and the founder who lends the money may be in control today but not tomorrow.

The first two mechanisms postpone compensation from one year into the next. If compensation moves into the following year, the employee's tax liability also moves into the following year. While deferring taxes is usually the preferred scenario, it may just be in the best interest of the company and the individual to take the tax hit early.

Consider the following example: On December 31, 2013, FelCo, Inc., has a cash balance of $20,000 in the bank. Having had a good year, FelCo decides to pay out its net earnings of $20,000 in management bonuses. However, the company needs operating cash for January because significant customer payments are expected no sooner than late January. Hence, going into 2014 with a zero balance in the bank would be highly risky. FelCo has two options:

- **In case FelCo prefers to pay taxes twice.** FelCo pays *corporate taxes* on $20,000 of 2013 income, approximately $7,000. On January 31, 2014, the company pays the executive team a total bonus of $13,000, which after tax withholding (assuming the 35 percent tax bracket applies) amounts to $8,450.

- **In case FelCo can afford to pay taxes early.** The company awards the bonus of $20,000 to the executives on December 31, 2013 ($13,000 after tax withholding). At the same time the executives grant back loans totaling $13,000 to the company. On January 31, 2014, the loans are paid back to the executive team, resulting in $13,076 (assuming a 7 percent annual interest rate) in cash compensation without any further tax liability.*

In both cases, the company had money to operate in January, but the net payment to the executives and founders is significantly better in the second case.

In venture-backed companies, a deferred compensation structure is rather difficult to establish. In fact, a new investor may require the company to eliminate all outstanding obligations to founders and employees as a condition of financial support of the company. In the best case, everybody gets paid the outstanding deferred compensation at the time of the next funding round (Thank you, ThingMagic investors!). In the worst case, key employees are asked to forgive the company all prior compensation commitments. Fortunately, hour and wage laws provide some protection for the startup team.

In the bootstrapped company, on the other hand, founders and executives are in control, and they can do what is best for the individuals and the company. Deferring compensation is a key cash flow management tool for a startup in bootstrapping or survival mode. Just make sure you double-check up-to-date federal and state compliance laws before you come up with a particular scheme.

COMPENSATION PACKAGES

Now that we have outlined the various equity tools, cash instruments, and other perks, let's look at what kind of compensation packages are needed to motivate people to work with and for you.

Compensation Packages for Founders

Before talking about money and stock, the founder team should try to answer a series of questions regarding the expectations of the indi-

---

* The transaction needs to be "at arm's length," that is, the loan needs to be legitimate so as to not upset the IRS.

viduals. The conversation should focus on the commitment of the individuals to the venture and the nature of the business they are about to launch:

- **Ask not what the company can do for you; ask what you can do for the company.** Which talent and level of commitment does the company need, and how long will it realistically take for the company to be successful? Are we going to run a lifestyle venture, or do we intend to cash out when the opportunity arises?

- **Any day jobs, anybody?** Is everybody planning on joining full-time?

- **Until death do us part?** Is the mission of the venture consistent with the individuals' long-term goals? Do the individuals anticipate leaving after a certain time period, or are they willing to stay on as long as it takes?

- **Passion is necessary and can't be bought.** Are all the individuals excited about the venture, or are some being dragged into it without really wanting to be there? Many a company cofounder recognizes an opportunity, gets on board, but then realizes that his heart isn't really in it. Needless to say, the absence of passion is not a good starting point for the individual or the company.

It is better to be up front about all these issues before working out an elaborate compensation scheme. When all the cards are on the table, a constructive solution can be found. If individuals are dishonest about their commitment, on the other hand, problems will show up quickly. Founders should consider writing down any understandings among themselves in a *founder's agreement*. A founder's agreement or letter of intent is typically nonbinding, but it helps avoid misunderstandings among the team and makes it easier to adjust plans, if someone on the team changes his mind.

Assuming that the founders are equally committed to the venture, equal stock ownership among the team members goes a long way in securing each other's commitment for the long term. Any deviation from that principle will cause feelings of injustice, underappreciation, and resentment, and it will result in a lack of motivation among the team members. In the exceptional case in which there have been truly disproportional contributions by a founder prior to others joining, an unequal distribution of founder equity may be justified. The default, however, should be equal ownership.

Differences in time commitment and early contributions can be justly accommodated through vesting schedules and cash compensation. If a founder can commit only 80 percent of his time to the venture, let him vest at 80 percent of the rate of a full-time founder and pay him 80 percent of the full-time salary. If any of the founders made significant contributions before the entity was established, they should be rewarded by declaring a percentage of their stock vested right from the start.

The vesting period for founder stock needs to reflect the realistic assessment of the time required to create a valuable company. Since building a business usually takes longer than one hopes, pick a vesting period on the long side, and complement it with a reasonably accelerated vesting clause. While six years of vesting may seem like an eternity for a young founder, developing a business stable enough to survive without its founders can easily take that long or longer. If the team is lucky and gets the job done more quickly, the accelerated vesting provision can take care of that fortunate but rare situation.

When we set up the ThingMagic restricted stock plan, we included a provision that allowed the company to buy back vested stock at the fair market value at the time of a founder's or executive's exit. While such clauses provide a significant retention incentive for the individuals, in practice the clauses cause more harm than good. Founders and employees are suspicious that the company will use the clause against them, when in reality the company won't even have the money to buy back the stock. The exceptions to this rule are closely held lifestyle ventures. For those companies, it does not make sense to let departing employees take their stock with them. ThingMagic dropped the buyback clause after the company converted to a venture-backed corporation.

As far as salaries are concerned, high-tech founding teams are at a greater risk of paying themselves too little than too much. The fact that young founders don't need much cash doesn't mean they don't deserve more. Teams who don't pay themselves at least within reach of market salaries are at risk of running out of steam before the company can sustain itself. Even founders right out of school grow up and develop needs. They will have families, and they will need significant monthly income. As you approach a funding round, also realize that the investors will not easily agree to a salary increase following the funding. The team more or less has to accept salary levels established prior to the funding event.*

As you come up with a compensation scheme for yourselves, be realistic, and make sure to not promise each other more than you can ultimately pay. A salary cut will be much harder to digest than an unexpected salary increase.

Compensation Packages for Advisors and Directors

Advisor positions in startups come in all shapes and forms, and so do the compensation packages that go along with these relationships. Unlike regular employment contracts, the regulators don't have much to say about appropriate compensation for advisory services. Hence, you can propose almost any arrangement you please. Here are a few typical advisor roles and suggestions for appropriate compensation models:

The Thesis Advisor. The young graduate admires, loves, or can't stand his thesis advisor.† Whichever emotion dominates the complex relationship between student and advisor, the student is heavily dependent on the goodwill of his mentor. When the two engage to found a company together, there is plenty of opportunity for controversy. The advisor, of course, should not exploit his position of strength,

---

* Similarly, investors are not likely to agree to changes in preestablished stock vesting schedules that would benefit the founders.

† I was lucky to be in the first category.

but even academics can have material desires. The student or former student should be prepared for uncomfortable conversations on the topics of how to involve, leverage, manage, and compensate his thesis advisor.

The discussion about contributions and compensation of the thesis advisor should be guided by the principle that compensation is awarded for services rendered and contributions made going forward. While pre-startup founding contributions need to be considered in coming up with a fair structure, the thesis advisor who thinks he should be compensated for his earlier inventions, fame, or patents alone has unrealistic expectations. Student founders are probably better off parting ways, rather than working with an overly entitled thesis advisor.

In most situations, the thesis advisor is best engaged as a member of the advisory board (see below). If his contributions are truly exceptional and time-consuming or if he joins the company full-time on a sabbatical, he should be treated as a cofounder of the company. In this case, the advisor-cofounder should be expected to take on some or all of the following responsibilities:

- Serve as executive chairperson

- Help with fundraising and business development

- Consult on technical matters

- Attend key strategic meetings

- Spend at least one full day per week on company business

If indeed the thesis advisor is contributing substantially in these roles, he deserves to receive the same equity stake as the full-time cofounders. If there is sufficient cash to go around, it may even make sense to pay him a reasonable consulting fee for the time spent working on behalf of the company.

The Independent Director. Whether or not the company is formally required to engage independent directors, a director can bring valuable perspective to the table, especially if "independent" means favorably inclined toward the founders and common stockholders. In bootstrapped companies, such outside eyes should validate the assumptions of the management team. In venture-funded companies, independent directors effectively act as mediators between the interests of common and preferred stockholders.

Independent directors should be people with experience in any or all of the following:

- Technical experience in the field of business

- Market knowledge and industry contacts

- Significant startup experience

- Connections in the venture funding community including angels, banks, and venture capitalists

Independent directors are expected to attend board meetings, stay up-to-date with the affairs of the company, and conduct business on behalf of the company when an opportunity presents itself. They do not just lend their names. Rather they are expected to contribute. In exchange, they receive between 0.05 and 2 percent of the company's equity depending on the stage of the company. Less than 0.5 percent of a young company is unlikely to get anyone to do anything. More than 1 percent would be unusual, unless the time contribution of the director much exceeds what can customarily be expected. As is true of any other equity stake, the director's shares or options should be tied to a vesting schedule of at least two years. Outside directors typically insist on directors and officers insurance as well as an indemnification agreement.

The Advisory Board. The members of the advisory board are compensated similarly to the independent directors, if not by quite as much.

The member of the advisory board needs to understand that he plays an important role in the structure of the company and that the company relies on him. The advisor should not be treated as an accidental addition to the staff. Rather, he fills a gap in the management team's experience, and he is expected to contribute specific industry insights, technical expertise, or contacts. Equity compensation (rarely more than 2 percent) along with potential consulting fees are provided in exchange for these contributions.

The Executive Advisor. An executive advisor distinguishes himself from an ordinary advisor by spending a well-defined amount of time with the company and by being available for day-to-day strategic and executive work. The relationship works best if the executive advisor gets assigned a specific role and responsibility in addition to being available for general questions and support. Executive advisors can be part-time acting VPs of sales, CEOs, VPs of business development, executive chairpersons, and so on. Setting up an executive advisor relationship provides an opportunity for the company and the advisor to get to know each other and figure out if a long-term, full-time relationship would make sense.

Compensation Packages for Early Key Executives

How can you attract experienced executives into a small company with little or no cash at hand? What appears to be an unsolvable problem is actually not that hard to do. The less the company has going for itself, the more potential upside there is. When you need to hire an executive who is prepared to work for a small technology company, you need to argue why the status quo is the perfect starting point for a big payday down the road. The executive has already made up his mind about the fundamental small-versus-large company trade-off. All you need to do is explain why you will make him rich.

In a bootstrapped venture, the salary for an executive can be way below market, but the equity award needs to be generous. The executive will argue that the risk in an unfunded venture is significantly higher than it is in a funded venture. The executive is right in that on average a funded company has slightly bigger chances of survival. However, he would be wrong to suggest that funded companies generate bigger returns for common stockholders. As Part II will show in detail, the payout for employee stockholders in bootstrapped companies that have not been diluted by preferred ownership can be significantly higher than it would be in venture-backed companies. So don't feel like you have to give away the house.

Venture-backed companies, on the other hand, are expected to pay executives close to market rate, but the equity grants tend to be much less generous. Unless you are trying to compensate a superstar, don't expect to give more than 4 percent of equity to anyone, except for the CEO.

The table below suggests reasonable stock grants to executives and key employees for bootstrapped, prefunding, high-tech startups and for venture-funded high-tech startups. The percentages can vary significantly based on the stage of the company and the importance of a role in the specific context.

	BOOTSTRAPPED STARTUP		VENTURE-BACKED STARTUP	
	Minimum	Maximum	Minimum	Maximum
CEO	4%	20%	2%	10%
VP sales and marketing	2%	8%	0.5%	4%
CFO	1%	5%	0.5%	3%
VP operations	0.5%	4%	0.2%	2%
Software or hardware architect	0.2%	3%	0.1%	1%
Senior engineer	0.1%	2%	0.05%	0.5%

Ultimately, you should think of the equity award to an executive in terms of the company's current stage rather than a fixed percentage. For example, if the funding round establishes a post-money valuation of $5 million, the equity award to a very senior executive could be expected to be in the $100,000 range—that is, about 2 percent of the outstanding equity.

Most executives cannot work for free, but some do not depend on a big monthly paycheck. As you evaluate candidates for employment, distinguish between financial flexibility and a lack of motivation to work hard. Don't hire executives who are planning on taking it a bit slower after some lucrative earlier startup years.

As was suggested above, try to establish a *flat* salary structure for the executive team if at all possible! When everybody makes the same money, the interests and motivations of the team are perfectly aligned toward maximizing the value of the equity down the road.

And How Much Are You Telling?

As you are trying to convince capable talent to join your company, be prepared to provide key data, including the number of shares outstanding, the vesting schedule, and the expected dilution in upcoming funding rounds. You should enable the employees to get a good picture of the state of affairs. The more you tell them, the more they will feel like they are part of the team working toward a common goal. Also, the more they know, the more they will understand when things go differently from the plan.

If there is information you'd rather not disclose, be explicit about that too. As long as you have a good reason, the employees will understand. After all, you both have the best interest of the company in mind.

LET THERE BE JUSTICE, IF YOU CAN AFFORD IT

Lesson 1: Allocate your founder stock early and generously, but make sure the vesting period is long enough for the company to reach stability and success.

Lesson 2: Neither under- nor overpay the founding team! Offer salaries that are generous enough so that founders and early employees can afford to work for your startup. At the same time, be reasonably stingy so that the company can stay independent of outside investors as long as possible.

Lesson 3: Flat salaries along with generous equity awards will align management around the common goal of maximizing the company's value.

Lesson 4: Don't play games with your employees! Let them know what their options and stock are worth and how the value is changing over time.

PART II

EQUITY FUNDING:
A DOUBLE-EDGED SWORD

CHAPTER 8

Venture Deals

Tary a little; there is something else.
This bond doth give thee here no jot of blood;
The words expressly are "a pound of flesh."
—WILLIAM SHAKESPEARE (1564–1616)
The Merchant of Venice

By the time we started looking for venture capital, we had boot-strapped ThingMagic for nearly five years. We had cultivated the unheard-of image of a profitable high-tech startup, and we had maintained the aura of profitability even after we departed from fiscal austerity and started to hire for growth. With revenue trending up and Walmart promoting a coming RFID revolution, it was rather easy to attract significant investments on very decent terms.

The VCs loved the idea of investing in a break-even company that was going to use capital to *grow* the business rather than build it in the first place. ThingMagic's founders, executives, and employees loved the fact that after the fundraising, the financial pressure would be off for a while. Little did we all understand at the time that we had irreversibly changed the culture of our company and that the invested money would eventually become an immense burden.

UNDERSTANDING PRIVATE EQUITY AND VENTURE CAPITAL

Venture capital firms are part of the larger category of private equity investors. (Figure 8.1 shows a graph of the annual number of U.S. venture capital deals and the total value of their investments.) They receive money from private individuals, wealth funds, and other institutional investors, and they invest the funds in high-risk private enterprises. A venture capital firm and its partners earn money through the following two mechanisms.

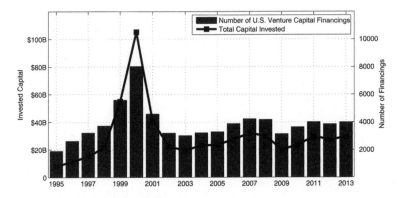

Figure 8.1: U.S. venture capital investments; annual number of VC deals and total money invested. Data source: PricewaterhouseCoopers and the National Venture Capital Association (NVCA), *Moneytree Report, Q1 1995–Q2 2013*, www.pwcmoneytree.com.

Each year the firm receives a percentage of the money under investment as the *management fee*, typically 1.5 to 2 percent. The management fee is used to pay for the operating expenses, salaries, and bonuses of the VC firm. While a VC partner doesn't get super rich on the management fee, she will earn an annual salary and bonus substantially above that of the technology executives the firm is investing in.

In addition, the partners in a VC firm participate in the gains of the funds they manage, the so-called *carried interest*, or *carry*. When a fund is liquidated after 5 to 10 years, the profit from the investment is split between the investors who put up the money (the limited partners) and the partners in the VC firm. The carry typically ranges from 20 to 30 percent of the gains. Some funds offer a minimum annual return for investors. Such minimum return is referred to as the *hurdle rate*, or *preferred return*, and it is typically in the 7 to 8 percent range. In rare cases, the carry is based on the gain of each individual deal, which absurdly favors the VC firm.

In summary, partners in venture capital firms have a very sweet deal. If they are "unsuccessful"—that is, they have managed their fund into a loss at the time of liquidation—they have earned "only" a few multiples of the annual salaries of typical technology executives. If they are very successful, they end up as billionaires.

On the flip side, partners in VC firms have to endure a firm's internal dynamics, which can be brutal. The stakes are high, and big egos are common. Partners managing a fund are typically compensated equally, not based on the performance of their individual portfolio. When a fund is not doing well, it is all but impossible to raise a new one and keep the firm going. Hence, the pressure on individual partners is enormous.

Venture capitalists do not enjoy a reputation of being the nicest professionals. Each time ThingMagic needed more money, I had to remind myself that VCs are not paid to be nice. They have been entrusted with somebody else's money. You, the entrepreneur, are going to spend that money, and the odds are rather small that the VC will ever get it back, let alone make a significant return. If the money is lost, the VC then has to explain the unfortunate outcome to disgruntled limited partners. Don't hold it against them if they are not giving you the velvet glove treatment, and certainly don't take it personally!

WHICH FIRMS TO SEEK OUT

ThingMagic's group of investors was rather diverse: a South American private equity fund, a hedge fund, a Hong Kong–based personal wealth fund, Taiwanese and U.S. corporate VC funds, early-stage VC funds, and private investors. Arguably, we had gathered an unmanageable group of VCs, and yet, strangely, the lack of coherence and divergent interests worked out well for us. Importantly, some of them had deep enough pockets to support us until the end. Others went out of business or lost interest, which for the portfolio company is actually not as bad as it sounds.

In figuring out which VC firm is best for your venture, first, get clarity on how much money you need to raise. There is no point in going to a late-stage fund if you are raising $100,000 to build a prototype. Likewise, you would be wasting your time asking an early-stage fund for tens of millions. There just wouldn't be that much money available, even if you managed to get the partners excited.

In sizing your investors, realize that in all likelihood, you will need more funding later. When the time for follow-up rounds comes, your existing investors are your best friends, and you'll want them to be sitting at the negotiating table with you. Think hard about what you need, and make sure that the firms you are talking to are in a position to fulfill those needs:

- **$100,000 to explore an idea.** *Seed-stage venture capital investments* are similar to angel investments except that the money is coming from a professionally managed fund. Seed funding is provided to small entities with limited capital needs to build a proof-of-concept, refine the business plan, hire early-stage personnel, or simply live a few months longer to get to the first real round of funding.

- **$1 million to get serious.** *Early-stage venture capital* is provided to complete commercialization of a product and start selling. Revenue is not technically required, but it sure helps. The startup is still at a stage where it can hide behind a large number of PowerPoint slides. But realize that the next time you are looking for funding, you better have some numbers that show a successful product rollout, a sizable revenue increase, or a trajectory toward breakeven.

- **$10 million to grow.** *Midstage or expansion venture capital* is provided to support and enable the growth of established industry players. Significant revenue, a solid product portfolio, and a name within the industry are expected. While you don't have to be profitable, you do need to be able to show a believable path to profitability. As a rule of thumb: determine whether your company could be cash flow positive if you gave up on the notion of growth. Don't worry, the VCs will want you to spend their money and grow. So just consider it a healthy thought experiment for your own benefit.

- **$100 million to get to the finish line.** *Late-stage, mezzanine*, or *bridge venture capital* is provided to companies in need of extra cash to get to an exit. Investors like to make a quick buck by investing a few dollars today and getting multiples on the investment back tomorrow. In reality, many such plans don't work out that way, and the mezzanine investment turns into a series of mezzanine investments. . . . Luckily, if you hook them once, they should be motivated to give you more later to protect the earlier investment!

Some firms create *thematic funds*, suggesting a preference for a particular type of portfolio company. The preferences range from a particular geographical focus (for example, emerging markets), to the type of technology supported (for example, enterprise software), to something as specific as iPhone apps. Thematic funds are good tools to bring some order to the madness of thousands of applications and hundreds of pitches VC firms are processing each month.

It comes as no surprise that venture capital firms cluster in very distinct regions—notably Silicon Valley, New England, and New York. The number of funded companies also clusters. About 30 percent of all U.S. financing deals go to Silicon Valley–based companies. Measured by the amount of money invested, the dominance of the Valley has increased even further over the last two decades: about 40 percent of all U.S. VC funds are invested in the Bay Area (Figure 8.2).[1]

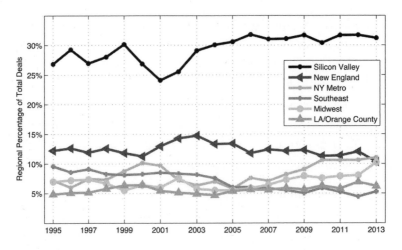

Figure 8.2: Percentage of U.S. venture capital deals by region: the six largest regions. Data source: PricewaterhouseCoopers and the National Venture Capital Association (NVCA), *Moneytree Report, Q1 1995–Q2 2013*, www.pwcmoneytree.com.

In recent years, the balance of power between startups and VC funds has shifted slightly in favor of tech ventures. This means better valuations, less dilution, and less severe liquidation preferences (Figures 8.3 and 8.4). The shift in negotiating power from VCs back to

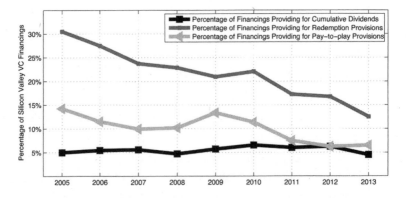

Figure 8.3: Use of cumulative dividends, mandatory or optional redemption provisions, and pay-to-play provisions in Silicon Valley financing rounds from 2005 to 2013. Data source: Fenwick & West LLP, *Trends in Terms of Venture Financings in Silicon Valley, Q2 2004–Q4 2013*, http://www.fenwick.com/publications/pages/default.aspx.

entrepreneurs, albeit modest, also helps entrepreneurs maintain control of their company. Google and Facebook are shining examples of tech ventures whose founders effectively maintained sole control all the way through their respective IPOs. Such entrepreneur-centric governance would have been completely out of the question even in the dot-com boom of the 1990s.

Should your fundraising efforts leave you unsuccessful and depressed, remember that the VCs depend on you as much as you depend on them, if not more so. Many an entrepreneur has built a successful company without taking a single dollar from a VC firm. However, no VC firm has ever succeeded without engaging with entrepreneurs.

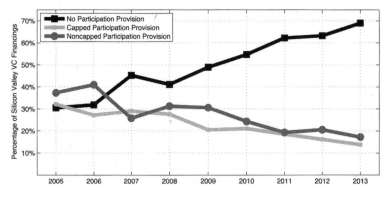

Figure 8.4: Preference provisions in Silicon Valley financings from 2005 to 2013. Fenwick & West LLP, *Trends in Terms of Venture Financings in Silicon Valley, Q2 2004–Q4 2013*, http://www.fenwick.com/publications/pages/default.aspx.

TERM SHEETS

Venture capital funds or investor groups express their willingness to invest in the form of a *term sheet*. The term sheet should be detailed enough so that both parties can get an unambiguous idea of the nature of the proposed transaction. At the same time, the term sheet should be as simple as possible so that neither party needs to spend serious lawyer money before it is even clear that there is a deal. Term sheets are almost always nonbinding. A term sheet can come in the form of a formal agreement, a letter agreement, or a simple e-mail.

The acceptance of the term sheet initiates the due-diligence period during which the investors get an opportunity to check on the company in detail and to make sure that there are no bodies buried in the cellar (see the section "Due Diligence and Closing" later in this chapter). The signed term sheet allows the startup to end the fundraising effort and shift back into normal operating mode.

In the following sample term sheet, only the key provisions are listed, and most of the legal boilerplate has been dropped:

LEACO, INC.
Summary Terms for Series A
Convertible Preferred Stock

This term sheet summarizes the terms proposed for an investment by a group of investors (the "Investors") lead by InvGroup, LLP (the "Lead Investor") in LeaCo, Inc., a Delaware corporation (the "Company").

1. **TYPE OF SECURITY:** Series A Convertible Preferred Stock

2. **AMOUNT OF INVESTMENT AND PRE-MONEY VALUATION:** The total Series A round shall amount to up to $5,000,000 at a price of $5.00 per share; the pre-money valuation amounts to $10,000,000, based on 2,000,000 fully diluted shares including an unallocated employee pool of at least 20% of the total. The Lead Investor will invest a total of $3,000,000, subject to the terms described herein, and it will serve as lead investor to negotiate transaction documents applicable to all Investors.

3. **DIVIDENDS:** Series A Preferred Stock is to participate in all dividends declared on an "as-converted" basis. No dividends are payable on Common Stock or any other Class of Preferred Stock without payment of similar and all accrued dividends to the Series A Preferred Stock.

4. **LIQUIDATION PREFERENCE:** The Series A Preferred Stock shall have liquidation preference over common shares equal to the price paid per share plus accrued but unpaid dividends. Any remaining proceeds shall be shared pro rata among common stockholders. A merger, consolidation, or similar event shall be treated as a liquidation event at the option of the Investors.

5. **CONVERSION:** The Series A Preferred Stock is converted on a one-for-one basis into Common Stock unless the conversion rate is subject to anti-dilution adjustment. Conversion of Series A Preferred Stock is mandatory on the closing of an underwritten public offering at an initial price to the public and a valuation of at least $25,000,000 and gross proceeds to the Company of at least $10,000,000.

6. **REDEMPTION:** The Investors will have the right to require that the Company redeem the Series A Preferred Stock at a redemption price equal to their respective per share purchase prices plus accrued and unpaid dividends of not less than 7% per annum. Redemption right may be exercised after year seven. At the option of the Company, the redemption price may be paid in three equal annual installments.

7. **ANTI-DILUTION:** The terms of the Series A Preferred Stock will contain standard "weighted average" anti-dilution protection with respect to the issuance of equity securities at a lower price in any subsequent round of financing.

8. **NEGATIVE COVENANTS:** The consent of holders of a majority of the Series A Preferred Stock shall be required for a merger, dissolution, sale of substantially all assets, dividends on common stock, amendments to certificate of incorporation, bylaws, or any changes in the rights of the Investors.

9. **NONBINDING NATURE:** The parties agree that this signed term sheet is nonbinding and subject to satisfactory completion of due diligence and successful negotiation and execution of definitive transaction documents.

Valuation and Dilution:
Do They Think So Little of Us?

When you receive the term sheet you have been waiting for so desperately, your eyes will quickly jump to the section on valuation. After all, how the stock is priced tells you how much value you have created, how badly your company shares will be diluted, and how highly the VCs think of you and your company. In reality, valuation is not necessarily the most consequential parameter of the financing deal.

For example, when investors put up money, they want to know that the cap table is balanced overall and that employees are adequately motivated to work at the startup. They may demand that a large percentage of stock gets reserved as an option pool for current and future employees (up to 30 percent of the outstanding stock is typical). This is a dilutive transaction for founders and existing common stockholders, but it is not reflected in the pre-money valuation.

What's more, the dilution of the existing stockholders and ownership transfer to the investors may be nonnegotiable from the VC's point of view. Professional investors tend to categorize investments and then price them with a particular ownership objective in mind. Venture capital investors like to be controlling, so they may just insist on a 51 percent stake post-funding.

Corporate investors, on the other hand, like to stay below a certain threshold so as to be able to keep the investment separate from the books and P&L of the corporation. Public companies, for example, are required, if the ownership stake exceeds 20 percent, to consolidate the P&L of both the corporation and the subsidiary. Also, many investors who are taking the lead want to see other investors get involved as well, as an endorsement of the company and as a source of additional funds.

If the investors have already set their mind on a specific target, there is not much use in pushing for a lesser number of shares. However, there may be an opening to negotiate for more money in return for the equity stake. In other words, if the investors are set on a particular ownership percentage, you can argue that more money is needed to manage the company to success. Using this strategy, you can effectively increase the pre-money and post-money valuations, and you can get more liquidity out of the deal.

Anti-dilution Protection: Much Ado About Nothing?

Reading ThingMagic's Series A term sheet for the first time, I was surprised and naïvely outraged to find that the investors would protect themselves, but not the founders or common stockholders, against future dilution. A few years later I realized that the anti-dilution clause didn't really matter all that much. It is one of those clauses that add a whole lot of anxiety to the process while not doing anyone much good.

Anti-dilution clauses kick in if the company raises a down-round later on—that is, if it sells future stock at a lesser price than previous rounds (Chapter 9). This situation, of course, occurs frequently, but when it does, the actual impact and enforceability of the provision are limited. Any new investor will likely insist that the anti-dilution clause be waived. The new guys at the table have two interests only. First, they want to own as big a chunk of the company as possible, and second, they want the founding and management team to stay ·motivated. The first goal is threatened if the anti-dilution protection gets invoked, since the earlier investors would effectively increase their equity position. The second goal is also threatened since additional ownership would be taken from the common stockholders, including the management team.

In Silicon Valley financings, about 93 percent of all deals stipulate *weighted anti-dilution protection*, which means the share price is partially adjusted to the price in the follow-up round.[2] In practice, the mechanism is implemented by adjusting the conversion rate of preferred stock to common stock. A typical formula reads as follows:

$$P_{A2} \quad = \quad P_{A1} \cdot \frac{(S_C + F_B/P_{A1})}{(S_C + F_B/P_B)}$$

where

P_{A2} = the new Series A conversion price

P_{A1} = the original Series A conversion price

S_C = the amount of common stock outstanding on an as-converted basis immediately prior to the Series B

F_B = the total funding amount in Series B

P_B = the Series B conversion price

If the clause does get invoked, the dilution effect is as much dependent on the difference in stock prices between the two funding rounds as it is dependent on the size of the new funding round. As long as the new round is small compared to the amount of outstanding common stock (on an as-converted basis), the dilution effect for the common stockholders and the new investors is marginal. If we start with the example of LeaCo's A Series (term sheet above) and add a hypothetical B round of $1 million at a price per share of $5 (half of the Series A price), the conversion price of the Series A preferred stock would change from $10.00 to $9.41 (effectively a 6 percent reduction in the Series A price).

About 4 percent of all venture deals stipulate full-ratchet anti-dilution protection. Under a full-ratchet provision, stock issued earlier is repriced at the new price. In this case, even a small amount of additional funding at a lower price will cause the cap table to change significantly in favor of the original investors. In the case of LeaCo, the effect of the small B round at half the earlier price would cause the A-round investors to double their equity. It would bring the total investor ownership to 55 percent, effectively handing control of the stockholder vote to the investors.

Fewer than 2 percent of all Silicon Valley financings have no anti-dilution provision at all; hence, *weighted anti-dilution protection* is probably the best you can get. If you are forced to raise a down-round later, raise as little money as possible to contain the dilution effect, or find investors with enough negotiating leverage to override the anti-dilution clause.[3]

Dividends: Really, Why Don't They Invest in Fortune 50 Value Stocks?

Dividends can have a major impact on the size of the preference pool years after the actual funding occurred. *As-declared dividends* are paid to preferred stockholders upon the decision of the board only, which is a rare occurrence in profit-less venture-backed companies. *Cumulative dividends*, on the other hand, automatically accrue each year, and the accrued amount is added to the liquidation preferences (see the section "Liquidation Preferences" that follows). See Figure 8.5.

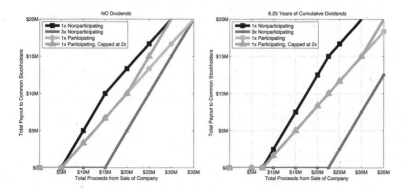

Figure 8.5: An example for payments to common stockholders in a liquidity event (acquisition) as a function of sale price and liquidation preferences. LeaCo: $5 million invested at a pre-money valuation of $10 million. Left: liquidation preference does not include accrued dividends. Right: liquidation event occurs six years and a quarter after the funding event. Preferences have increased to $7.5 million due to an 8 percent annual cumulative dividend rate. Liquidation preferences much above a 1X multiple have become uncommon in recent financings.

Dividend rates are typically in the 6 to 8 percent range, even when government-established interest rates are low. Venture capital firms argue for a high dividend rate in light of the riskiness of the investments. Of course, this is a circular argument because the risk premium is already built into the price of preferred shares.

While as-declared dividends are standard practice, cumulative dividends are unusual (Figure 8.3). Between 2004 and 2012, on average only about 6 percent of Silicon Valley financing rounds provided for cumulative dividends.[4]

Redemption: Don't They Know That Seven Years from Now, We'll Be Either Dead or Fabulously Rich?

Venture capital funds like to have the option to sell their investment back to the company at some time in the future. Redemption can happen after as little as five years, and it usually includes accrued dividends. The term either stipulates mandatory redemption or, more commonly, redemption at the discretion of the investors (Figure 8.3).

Unfortunately, redemption provisions in startup financings almost never result in money actually being returned to the investors. A few years after a financing round, the company is either doing well, and the investors have no interest whatsoever in taking their money back, or the company is doing badly, in which case there won't be any cash to redeem the stock. At the same time, no new investor would be interested in buying out the old ones.

Over the past nine years, about 25 percent of Silicon Valley venture deals included a redemption provision, with a steady decline from 30 percent in 2004 to about 20 percent in 2012. If your investors insist on a redemption provision, negotiate for as long a redemption period as possible.[5]

Negative Covenants: They Clearly Want to Be in Control, Don't They?

Negative covenant provisions stipulate the right of the investors to veto major transactions by the company, even if those transactions have been sanctioned by the board of directors. The *veto right* is also called the *investor block*. From the point of view of the investors, negative covenants are the fallback control mechanisms in portfolio companies for which they cannot control the stockholder vote or the board vote. Major transactions covered by the covenants typically include merger transactions, the sale of substantially all assets, acquisitions, future funding events (including bank loans), amendments to the bylaws, and changes to the investor rights.

Negative covenant provisions are meant to prevent executives and future investors from ganging up against the current investors. Yet the provisions can also cause investors to hurt each other. A diverse investor group can find itself deadlocked, unable to find a majority for any reasonable action or decision.

When ThingMagic needed yet another round of funding after a series of capital injections, the representative for the controlling investor would not agree to a new fundraising effort. His concern was to get his money back as quickly as possible. He paralyzed the company against all good judgment, much to the disbelief and financial detriment of his fellow investors. Approval of future fundraising is perhaps the most troubling among the restricted transactions. Most investors can agree to a lucrative liquidity transaction, but they don't like the prospect of dilution and loss of investor rights to a new investor group.

Negative covenants are an integral element of the VC establishment. As an entrepreneur, you will not be able to negotiate them away anytime soon. However, you can minimize the impact on common stockholders:

- Negotiate for a **clear definition** of the types of transactions included in the provision.

- Establish a **minimum transaction value**, below which no approval is required.

- Argue that the investor vote should be based on a **simple majority** so as to not paralyze the company. Supermajorities are harder to get.

- **Carefully manage the preferred stock cap table.** The specifics of the investors' holdings determine the investor vote. You need sufficient votes aligned with executives and the board, but you do not need everybody's approval to govern the company.

LIQUIDATION PREFERENCES: THE BIGGEST NUISANCE OF THEM ALL

Preference provisions are easily the most important clauses on the term sheet. Once you grant investors the right to get their invested millions back before you and your fellow startup employees get a dime, the financial and social dynamic of your venture is forever changed. When we negotiated ThingMagic's A round, we did not remotely think that the preference provision would ever come into play. I suspect our expectations in the matter were not any different from most entrepreneurs: why would the company's valuation ever drop? After all, we had just secured a hell of a lot of money to grow the company.

Liquidation preferences specify how proceeds from a liquidation event are distributed among preferred and common stockholders. Investors are typically issued convertible preferred stock. At the time of a liquidity event, the investors have the right to convert their preferred shares to common shares and to be treated like every other common stockholder. If the investors do not convert, the liquidation preferences take effect.

Liquidation preferences exist in three variants (see also Figure 8.5). Let's consider LeaCo's hypothetical Series A round, which we will then use to explain the impact of the three preference types:

- Pre-money valuation (only common stock outstanding): $10 million

- Capital invested (preferred stock): $5 million

- One-to-one conversion between preferred and common stock

Nonparticipating Liquidation Preference

This one is easy. Depending on the multiple of the preferences (1X, 2X, 3X), the investors get 1, 2, 3 times their money back before the remaining proceeds (if there is anything left) are distributed to the common stockholders. If LeaCo's term sheet stipulated a nonparticipating 1X liquidation preference, a liquidity event would result in the following payouts:

- If LeaCo sells for less than $5 million, the investors get all the money.

- If LeaCo sells for between $5 million and $15 million, the investors get $5 million, and the common stockholders get the rest. This range is sometimes referred to as the "zone of indifference."

- If LeaCo sells for more than $15 million, the investors will convert their shares to common shares, and everybody will get their pro rata shares. Given that the investors own exactly one-third of LeaCo, they will receive exactly one-third of the money.

Participating Preferred Liquidation Preference

In this case, investors get their money back first (much as in the preceding case), but they are also participating in the distribution of any proceeds beyond the initial preference. If the preference is 1X, the investors get their money back, and they also get the pro rata share of any distribution of proceeds beyond the preference (on an as-converted basis). If the term sheet for LeaCo had a participating preferred clause in addition to the 1X preference, the distribution of proceeds would look like this:

- If LeaCo sells for less than $5 million, the investors get all the money.

- If LeaCo sells for more than $5 million, the investors get $5 million, and the investors and common stockholders share the rest on an as-converted basis. Given the investor ownership of one-third of the company, the investors get $5 million and one-third of all the proceeds in excess of $5 million.

Capped Participating Preferred

The *capped participating preferred* approach specifies two multiples. The first multiple clarifies the actual liquidation preference, and the second multiple specifies until which multiple of the invested money the investors are entitled to a participation in the proceeds. For example, a 1X participating liquidation preference capped at 2X means that investors get to participate until they have recovered 2X their investment. How do LeaCo's stockholders fare in this case?

- If LeaCo sells for less than $5 million, the investors get all the money.

- If LeaCo sells for between $5 million and $20 million, the investors get $5 million, and they share the rest with the common stockholders on an as-converted basis. Given that the investors own exactly one-third of LeaCo, they will receive one-third of the proceeds in excess of $5 million. For example, at an $11 million sale price, the investors receive a total of $7 million, and the common stockholders receive $4 million.

- If LeaCo sells for between $20 million and $30 million, the investors get $10 million, and the common stockholders get the rest.

- If LeaCo sells for $30 million or more, the investors will convert their stock to common. Both investors and common stockholders will get their pro rata share of the total sales price.

Liquidation preferences may cover the money invested plus an annual dividend. Assuming an 8 percent cumulative dividend for LeaCo above, and assuming that LeaCo is being held a little over six years before it is sold, the liquidation preference now covers 150 percent of the money invested. In other words, LeaCo's employees and management team receive cash for their common stock if proceeds from the sale of the company exceed $7.5 million to $22.5 million depending on the liquidation multiple.

Figure 8.5 illustrates the total payout to common stockholders as a function of sales price and timing of the liquidity event (without dividend payments or after 6.25 years of dividend payments).

Liquidation preferences are meant to protect the investors on the downside. However, that protection can backfire and cause both preferred and common stockholders to fare badly. Given the way most clauses are written, certain exit scenarios result in a flat return for common or preferred stockholders. Most important, if a venture is liquidating *in the preferences*, the common stock is worthless, and the employees and executives do not benefit from the company sale at all. Anticipating such an outcome, many executives might not feel too motivated to work hard on the investors' behalf. Rather, they might start looking for another job and jump ship before the transaction closes.

Likewise, in the case of nonparticipating liquidation preferences, there is a range of exit values for which the specific outcome is immaterial for the investors. Will the investors spend time and energy to negotiate hard on behalf of the company if they know that it won't make one bit of difference to their bottom line? It is arguable that common stockholders are actually better off with a structure that includes noncapped *preferred participation* since it avoids any zone of indifference and investors are motivated to maximize the deal value irrespective of sale price.

In recent Silicon Valley financing rounds, on average 45 percent of financing events included no *participation* clauses, 24 percent included *capped participation*, and 29 percent included *noncapped participation* (Figure 8.4). So, don't get your hopes up! Every institutional financing round stipulates liquidation preferences.[6] All you can negotiate for are decent terms consistent with or better than industry standard. Fortunately, in recent years, multiples in excess of 1X have become very uncommon.

Senior Liquidation Preferences: How Much Worse Can It Get?

The preference situation gets complicated after multiple financing rounds. New investors will push for their liquidation preferences to be *senior* relative to earlier investments. Previous investors, on the other hand, will push for their investment to be on equal footing with the new preferences (*pari passu preferences*). Who wins depends on the leverage of the old versus the new investors and on the respective investment philosophies.

Senior liquidation preferences in a term sheet offered by a new investor group present a leverage point for the management team to negotiate better terms with existing investors. Lest the existing investors want their ownership to be crammed down, the existing investors should maintain control of the funding process and offer better terms than the new guys.

About 42 percent of recent post–Series A funding events in Silicon Valley stipulated senior preferences. Out of those funding rounds that stipulated senior preferences, about 79 percent called for multiples between 1X and 2X, about 16 percent called for multiples between 2X and 3X, and about 5 percent called for multiples in excess of 3X.[7]

Executives and common stockholders should push investors to institute a *pay-to-play provision* in the preferred stock agreements. The clause requires investors to invest in future rounds on a pro rata basis in order to maintain the preferred status of their investment. If investors are unwilling or unable to put up additional funds in the future, their stock converts to common, reducing the overall preference burden on the venture. Due to this provision, ThingMagic surprisingly came out of certain future funding rounds with fewer preferences than going into them.*

---

* This usually works only once, when a minority of the preferred stockholders doesn't "pay" and the majority of investors votes to proceed with the minority's conversion. If a provision was not negotiated at the time of the original financing, the company should consult with counsel about implementing a retroactive pay-to-play provision leading into the next round.

In a variation of pay-to-play provisions, investors may try to institute shadow preferences. Shares of investors who are no longer investing are converted to *shadow preferred stock*, rather than to common stock. Shadow preferred stock enjoys liquidation preferences over common stock but is junior to new preferred stock. In many ways this is the worst outcome for the common stockholders!

DUE DILIGENCE AND CLOSING

Closing on a financing round or an M&A transaction is a stressful undertaking. Indeed, things can go south until the very last minute. ThingMagic was peering into the financial abyss on the occasion of its A round, leading up to most of its follow-up funding events, and again leading up to its acquisition. Luckily, we were able to close a deal at the last moment in every one of those dire situations. Plenty of other tech ventures are on track to close on a lifesaving financial transaction one day, but instead they end up in ruin and bankruptcy the next.

Riches or Rags?

The activities in the weeks before closing on a major financial transaction can put even a healthy venture in a very vulnerable position. Much of the company's effort is spent on the closing, while other options are no longer being pursued. There simply aren't enough resources and bandwidth in a small venture to work on multiple deals at the same time.

When the potential investors or buyers smell desperation, the temptation is great to exploit the venture's vulnerability. As buyers or investors threaten to walk away from the deal at a late stage, terms may find their way into closing documents that no self-respecting founder or CEO would ever have considered signing.

What can you do to avoid last-minute pressure?

- **Watch out for your timing.** Don't wait until the last minute to start raising money. Closing on any financial transaction takes time, and it always takes longer than you think. Schedule your effort in such a way that you have enough run rate left to make it through a second fundraising cycle if necessary.

- **Avoid exclusivity (no-shop) provisions.** Investors don't like it when an opportunity is being snatched up by someone else at the last minute. However, from the venture's point of view, the promise of exclusivity severely limits the deal options and kills every bit of leverage the venture had. If you can't avoid the clause altogether, make sure that there is a time limit on the exclusivity arrangement that allows you to pursue other options if the deal does not go through as scheduled.

- **Maintain a Plan B.** If there is no Plan B, pretend there is one as best you can. You want to represent to the world that your company will live, whether or not the particular deal goes through. If you don't believe it, nobody else will.

Discovery and Disclosures

The focus of the due-diligence process should not be the valuation of the company or its future prospects. In theory those questions have already been answered leading up to the term sheet. Rather, the investors are being given the opportunity to confirm that everything they were told by the company is actually true. If the process uncovers any fact that materially affects the value of the business, it is legitimate to renegotiate. If there aren't any surprises, the deal should go through as stipulated in the term sheet.

The due-diligence process includes two key elements. First, as part of the *discovery process*, the company delivers every document relevant to the state of the business to the investors or buyers. Second, as part of the *disclosure process*, the company issues statements in response to the investors' questions, testifying on the state of the business.

Being well organized in the due-diligence process is as good a display of the company's strength as the actual data produced. The other party will appreciate a smooth process where time and money are being spent debating the content rather than the administration of the process itself.

Don't sign the term sheet and start randomly sending e-mails with attachments. Take the time to organize the information, and ask the other party to respect the fact that you want to go through the due-diligence process in a systematic fashion. Rather than cluttering inboxes, set up a secure website from which the various documents can be downloaded. There are services out there to provide portal functionality for exactly that purpose.

It is perfectly fine to suggest that a particular document or a piece of information is not relevant in your case. You will be getting a boilerplate list of questions. The investors may not adjust their request for information to your specific company, but instead, they will present a comprehensive template that makes sure no potential issue is forgotten. Hence, don't be shy in pushing back on certain documents you deem irrelevant. If the investors disagree, they will let you know.

So, you might ask, what do you do if there ARE dead bodies in the cellar? After all, that is what the whole process is supposed to uncover. If you suspect that an issue will truly threaten the outcome of the negotiation, don't wait until the last minute to bring it up. An investor really can't argue for a change of terms based on an issue that was disclosed and made explicit prior to signing the term sheet. However, after the term sheet has been signed, any discovery is a welcome opportunity for the investors to renegotiate.

The due-diligence process has never caused investors to improve on the deal terms. You just don't hear: "I wasn't aware that your customer base was so diversified and strong. Based on that, we would like to increase the pre-money valuation by 20 percent." Given that reality, make sure you have all the materially good information and all the potentially damaging information on the table when you decide to accept an offer. The former avoids the acceptance of an inferior offer, while the latter avoids a negotiation to the worse during the discovery period.

If you know of an issue that you think will likely cause a problem with the investors, choose your words and documentation carefully.

Your obligation is to disclose the facts. An interpretation of the facts is not required, and it is a matter of judgment. If color commentary doesn't help, don't provide it! On the other hand, if you can spin the situation in a good way, be as verbose as you deem helpful to the cause!

Avoiding Last-Minute Deal Killers

As the closing is approaching, prepare yourself for emotions to run out of control! As the deal looks more and more inevitable, the prospect of its falling apart looks more and more catastrophic. Once you've prepared yourself mentally and financially for the peace and security of a few well-funded months, the thought of hitting the road again fundraising or going bankrupt certainly is stressful. This dependency makes the company vulnerable to last-minute hardball negotiation by the other party.

The need to close also presents an opportunity for internal stakeholders to negotiate for some benefit as a condition of staying put and cooperating. All of a sudden, your VP of development wants a bigger option award, the VP of sales wants a bigger commission, and the controller wants to be promoted to CFO. No matter how good the deal is for the company, certain individuals will think of some request to improve their personal financial and career situation.

Some of the internal bickering can be avoided by simply not telling anyone who doesn't have to be involved. Everybody should be informed about the deal before the public announcement, but there is no need to tell rank-and-file employees about the money coming in before the deal is signed.

For those employees you have to tell beforehand, present the deal as nonnegotiable. Use any excuse you can come up with including, "I will present your request to the board, but I don't think they'll approve it," "If you insist on this issue, the deal is dead," or "The investors will not go for this."

There comes a moment in every major startup transaction when suddenly the outcome and closing depend on the willingness of a small number of individuals to make a significant personal sacrifice.

If at that juncture you don't observe any F-bombs flying, the deal was probably immaterial in the first place. From the founders' point of view, the most likely points of contention include *revesting of founder stock*, *multiple preferences*, *option pools* that dilute the common stockholders only, and above all, *noncompete clauses* for key employees. Understandably, investors don't like it when the people they are investing in go off and work for the competition. Also understandably, founders don't like to be restricted in their future careers.

What can you do to avoid a showdown around noncompete clauses and other critical founder rights? The investors may not bring up such personnel issues early on, but you as a founder should. If you want to make sure that *nonfinancial terms* don't become a stumbling block at the last moment, make sure to discuss your concerns prior to signing the term sheet. Ideally you hold a draft of the noncompete agreement in your hands before you make your personal commitment.

Beyond emotions and civil liberty concerns, certain legitimate deal killers may come up at the last minute, including the following.

Legal Hostility and Lawsuits. If you are being sued during the due-diligence process, expect to close at greatly diminished terms or to not close at all. Therefore, don't provoke your archenemies during fundraising time. If you are being humiliated, swallow your anger until you have the financial backing to start a real war. Under no circumstances should you get involved in any kind of legal dispute prior to closing.

IP Threats and Saber Rattling. IP claims by a third party are another popular reason for nervous investors to bail out. Patent trolls like to attack at critical transition times of a young and promising company. For example, when a company files for an IPO, it is basically asking the world to come forward with claims anyone might have. Justified or not, the company will consider settling just to keep the issue off the table. In the case of a private financing, the best defense is secrecy. Once the deal closes, the leverage of the patent troll is diminished.

If you are in the unfortunate position to get that letter from an inventor or organized patent troll, keep your cool, and position the situation to the investors rationally. Patent holders send out letters as a matter of due course. Once they have informed you about the patent, you cannot claim ignorance should you ever be found to be infringing. However, these actions say nothing about the validity of the patent or the infringement of your products. Furthermore, patent holders announce their IP broadly. They almost never state a particular reason why your product might be infringing.

The likelihood that anything will come of the initial notification is very small. As a last resort and only if absolutely necessary, get your IP lawyer to draft an opinion of noninfringement. This costs money, but, statistically speaking, chances are good that you can argue for the invalidity of the patent or noninfringement of your product.

Deteriorating Financial Situation. By all means, avoid any action that suggests the company's financial situation or prospects changed materially in the days and weeks before the closing. Keep your best customers happy during the fundraising process so as to not threaten your revenue outlook! Be nice to your vendors so that you won't have to look for new partners or pay more for services and products before you close on the financing round.

Regulatory Compliance. A bad gut feeling does not require disclosure. A conversation with an expert in regulatory compliance, on the other hand, does, and so do negative test results, whether they were obtained in-house or by a third-party lab. Make it clear to your tech team that any regulatory setback could have a catastrophic impact on your fundraising efforts!

As you approach closing, you will be busy working out the legal documents, and you won't have the time to follow up on every single tactical issue, big or small. That's a good thing! What you don't know, you don't have to mention in the closing documents.

TAKING A BITE FROM THE APPLE

Lesson 1: Negative covenants make the investors the most powerful stakeholders in the venture-funded startup. You cannot avoid them, but you can limit the damage by carefully managing the investor dynamic and ownership percentages.

Lesson 2: Liquidation preferences are a nonnegotiable reality of venture financing. However, you can and should negotiate hard on the type of preference to be applied. Try to align the interests of common stockholders with the interests of either existing, new, or future investors.

Lesson 3: Insist on a pay-to-play provision in the investor agreement. It will allow you to get rid of the preferences awarded to investors who can't support you in the future.

Lesson 4: Maintain a Plan B until the moment the deal closes. Without an alternative, you are minimizing your negotiating leverage, and you are exposing your company to the possibility of a meltdown if the financing falls through at the last minute. Never ever count on things to go smoothly!

CHAPTER 9

Raising a Down-Round

Much good work is lost for lack of a little more.
—EDWARD H. HARRIMAN (1848–1909)

If you are raising money at a pre-money valuation below the post-money valuation of the last funding event, you've got yourself a *down-round*. To state it differently, your new investors are paying less per share of stock than your previous investors paid (Figure 9.1). After ThingMagic's spectacular A-round valuation, our stock price decreased in every single follow-up financing event. We clearly had overestimated the prospects of our company and our industry, but so had all the other stakeholders including the investors, analysts, and the general public!

A down-round hurts all stakeholders: existing investors are forced to write down their investment; employees now work for stock options valued below their strike price; executives who purchased restricted stock wonder whether they should consider the investment a sunk cost and move on to the next opportunity.

On the bright side . . . if you are about to raise a down-round, you must have gotten a really good deal in the previous funding event. Your company hasn't fulfilled its promise yet, but you did one hell of a job selling it to your investors the last time around.

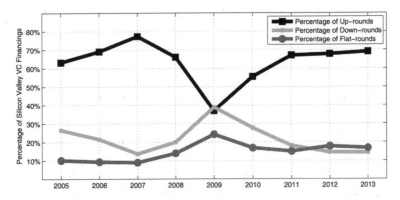

Figure 9.1: Percentage of up-rounds, flat-rounds, and down-rounds among Silicon Valley post–Series A financings, 2004 to 2013. Data source: Fenwick & West LLP, *Trends in Terms of Venture Financings in Silicon Valley, Q4 2004–Q4 2013,* http://www.fenwick.com/publications/pages/default.aspx.

GOING BACK TO THE WELL

If your venture is hurting and you need more money, first go back to your original investors and ask them for more. Investors don't like throwing good money after bad, but they also don't like portfolio companies going out of business or being strong-armed by new investors. Given the alternatives of (a) the portfolio company going bankrupt or (b) the new investors putting in funds at onerous terms, there is a good chance that they will help you out.

Existing investors like to be part of future funding rounds lest their interests will be butchered by new investors. When a new investor group does come in, they try to negotiate a lot of the terms away that were meant to protect the existing investors. The new investors drastically cut the perks assembled in the previous round and then reestablish them for their own benefit. The previous investment agreement likely included anti-dilution protection, but the new investors won't want to honor the provision! The old investors held liquidation preferences, but the new investors will make sure that their own liquidation preferences are the most senior of them all. Board seats naturally are transferred to the new guys.

You may ask, what happens if the existing investors don't agree? Well then, there won't be any new money! In a down-round scenario, the guy who comes in last with a lot of liquidity always holds the best cards (Figure 9.2).

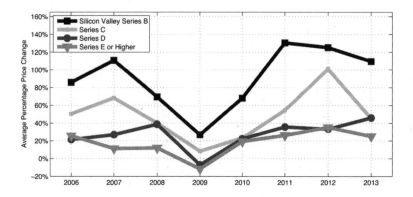

Figure 9.2: Share price change compared to the previous round in Silicon Valley post–Series A financings by type of series, 2006 to 2013. Data source: Fenwick & West LLP, *Trends in Terms of Venture Financings in Silicon Valley, Q1 2006–Q4 2013*, http://www.fenwick.com/publications/pages/default.aspx.

Management and common stockholders can use these power games among the investors to their advantage. Management is trying to minimize dilution and liquidation preferences as much as possible, especially when things are tough going. Both objectives are easier achieved with the existing investor group than with a bunch of new sharks, who are smelling blood in the water. The existing investors already have significant skin in the game. They are reluctant to negotiate hard on the terms in the new round because doing so will effectively hurt their earlier investment.

Repricing the Last Round, or How to Make the Down-Round Go Away

When ThingMagic needed more cash at the worst possible moment—that is, in the midst of the meltdown of our entire industry—

our CEO had the brilliant idea to *reopen* our A round. Rather than raise a B round at a very unfortunate new stock price, we repriced our A round and offered additional stock at a blended price. We pretended that the round had never closed and that we were still raising money against the structure of the last investor agreement. Don't get your hopes up too high! This is not a silver bullet to keep your share price where it was before. It does help contain the damage though.

Here is how the mechanism works by way of example. Let's assume the hypothetical CasCo Inc. raised $5 million in A-round funding against a pre-money valuation of $10 million and at a price per share of $10. At the time the investors received 500,000 shares of stock, bringing the total of outstanding stock to 1.5 million shares. After the A round, the investors effectively owned 33 percent of CasCo.

The company spent all the cash over the course of a year, but none of the anticipated revenue milestones were met, and the company did not reach profitability. The investors are willing to put in an additional $1 million to keep CasCo alive, but they are set on a new pre-money valuation of $5 million.

If CasCo accepted the funding as part of a B round, the new stock (300,000 shares) would be issued at $3.33, bringing the post-money valuation of the company to $6 million. The investors would then own 44 percent of the company. They would have to write down their earlier investment by 66 percent, bringing the book value of the investment from $6 million down to $2.66 million. Not pretty! Furthermore, previously issued employee stock options would be significantly under water, prompting uncomfortable conversations at the watercooler.

The alternative approach is for CasCo to reprice the A round and accept the additional $1 million as a second A-round closing. The repricing is calculated based on the notion that the investors will end up with the exact same ownership percentage as in the B-round scenario. In other words, the new pre-money valuation becomes the basis of the repricing of previously issued stock. Applying this principle, the stock price is set at $7.50, resulting in a post-money valuation of $13.5

million. Thus 44 percent of the company is owned by the investors. The number of outstanding shares (1,800,000) and the number of shares owned by the investors (800,000) are also the same as in the B-round scenario.

The effective drop in stock price from $10 to $7.50 is much easier to digest than the dramatic slide caused by the B round. CasCo should be able to rationalize a 25 percent drop in share price to its employees and other important stakeholders. The investors have more degrees of freedom in setting the book value of their investment.

In case you want to try this at home, here is the formula to reprice a round based on the new negotiated pre-money valuation and a number of other parameters:

$$SP_r = \frac{I_{1st} + I_{2nd}}{\frac{I_{1st}}{SP_{1st}} + \frac{I_{2nd}}{SP_{2nd}}}$$

where

$$SP_r = \text{repriced stock price}$$
$$SP_{1st} = \text{negotiated stock price initial investment}$$
$$SP_{2nd} = \text{negotiated stock price second investment}$$
$$I_{1st} = \text{amount invested in initial investment}$$
$$I_{2nd} = \text{amount invested in second investment}$$

In addition to the benefits already mentioned, there are further advantages of repricing a round:

- **Sorry lawyers, no six-figure fee this time around.** Administratively, you save yourself a ton of work and expense. Rather than asking a lawyer to draft up a complete set of new funding documents at a time when money is definitely short, you simply change some numbers in the previously agreed-upon documents (a somewhat simplistic statement, but largely true).

- **There's a nice and simple ownership structure.** The capitalization table of the company stays simple because no new class of stock is introduced. A few numbers change, but no new financial instruments are added. Future funding or liquidity events will be easier to administer based on a clean cap table.

- **There's no down-round stigma.** The negative connotation of the down-round event is all but eliminated from the history of the startup. Of course, insiders and stockholders will know about the *cover-up*. As far as everybody else is concerned, there was just one round of funding and one share price.

- **The stock price dipped, but it did not collapse.** The price per share of stock is kept close to the former price, avoiding all kinds of problems with regard to option pricing, restricted stock, and future funding rounds.

And how does the government feel about such a deal? As a private company, most of this is between you and your investors. The IRS is very concerned that you don't value newly issued incentive stock and options too cheaply. Make sure to not do that, and have an accountant check out any deal you come up with.

Bridge Loans: Hoping for Things to Get Better After All

Bridge loans by the existing investor group may be the only viable savior when the venture needs more money but has no time to go fundraising. Bridge loans are stopgap solutions, leading up to a bigger round or a liquidity event. They are particularly effective when it is not yet clear where a venture is going with its funding strategy.

Most bridge loans are structured as convertible debt. As soon as new liquidity is generated with a well-defined share price, the loan converts to equity, typically at a discount. The mechanism used is similar to convertible loans issued to very early stage companies. At a time when it is impossible to put a price on equity, investors and management prefer to *punt* the issue until such time that a funding event specifies a new share price.

Before you ask your investors for a convertible bridge loan, think hard about whether you expect the price per share in the next round to be higher or lower than the price in the last round. If you expect the price per share to go up, you should ask for a bridge loan with a conversion discount, typically in the 15 to 40 percent range.* If you think the share price is more likely to fall, you are better off selling straightforward equity based on the price set in the last round. Your investors, of course, may have their own opinion about what they are and are not willing to do for you.

Depending on how the next round of funding does come about, the terms of the bridge funding may be altered, and the discount may be negotiated away or absorbed in the overall arithmetic of the new deal. Especially with a new set of VCs, there is no telling what sacrifices they will ask from the common stockholders and the existing investors.

B, C, D, E, F . . . ROUNDS

Alien Technology in Morgan Hills, California, was one of Thing-Magic's fiercest competitors for many years. The company had already raised and spent a boatload of money by the time it entered the RFID market in 2001. When Alien tried to launch an IPO in 2006, it had spent more than $250 million. The IPO failed, but the fundraising in the private market has continued ever since. To date, Alien has raised in excess of $350 million from private investors, which makes it a fascinating case study for successful capital campaigns. The company has been able to continuously raise venture capital, even after it has become clear that it will never be able to return the invested money, let alone pay the collective investors a multiple on their investment.[1]

How is this possible? Why would investors put more money into a company that is worth a fraction of the money previously invested? The answer lies in the powerful magic of liquidation preferences. Investors agree to invest, but only under the condition of a multifold liquidation preference. The new shares are considered senior not only relative to the common stock but also relative to the previously issued

---

* In some cases discount ranges are set to increase over time until conversion occurs, and they can graduate to as much as 50 percent.

preferred shares. In other words, the new investors will get a multiple of their investment from a liquidity event, before anybody else gets anything . . . if the company sells before new funds are needed, that is!

For example: the hypothetical AnaCo Inc. raises a J round of $10 million. The investor agreement stipulates a 2X senior liquidation preference. If the company sells for $20 million before any more money is raised, the J-round investors just doubled their money. Nobody else gets a penny, including those investors who believed in the very same story when they invested in AnaCo's B, C, D, . . . and I rounds. Get the idea?

Terms: Beggars Can't Be Choosers!

If you are looking for a new group of investors while your venture is doing badly, your negotiating leverage is weak at best. First, you have to find someone who still believes in the story of your technology and your market. If you succeed in convincing investors that you have it in you to turn your company around, you can be certain that they will agree to invest only if the following conditions are met:

- **We, the new guys with money, are in control now.** The new investors take over as lead investors and control the investor vote. The new guys will get more shares for a lesser price than previous investors.

- **We dilute the old guys as much as we can.** Anti-dilution provisions from previous funding rounds will be invalidated or overwritten.

- **As a matter of fact, we don't think the previous investors are entitled to anything.** Liquidation privileges held by previous investors are being greatly reduced, if not eliminated. The new preferred stock will carry a more comfortable and more senior liquidation preference than any stock issued previously.

The new investors will focus on reducing the rights of previous investors who are no longer participating. As far as the new team is concerned, those older guys are no longer good for anything. Their money has been spent, and they are not forthcoming with new funds. So why give them more than absolutely necessary to get them to sign the new funding documents.

In most cases, not all existing preferred stockholders have to agree to the new funding round. It suffices for the management and new investors to get enough support from the old investors to prevent an investor veto on the funding round. In reality, investors who are no longer participating have really no choice but to approve new money coming in, if it can be found. Given the alternative of bankruptcy or foreclosure, rational investors would allow the venture to stay alive.

The Fate of Common Stockholders

As you go about raising money in a late-stage round, both old and new investors have an interest that the employees and executives of the company are *being taken care of*. If the company has no means to motivate key talent, current key performers will pursue opportunities elsewhere, and job candidates will not even come to a second interview. A structure that neglects to offer employees good incentives and upside will not work, and it will ultimately destroy what hope there is left for the preferred stockholders to make any money.

What can be done to protect common stockholders? Your concern should be current and future employees. Former employees and common stockholders who got rewarded with equity years ago are no longer there and won't help to get things back on track. The employees who stayed are the ones who need to be protected from the dilutive effects of a down-round:

- **Issue generous equity grants to long-time employees.** The new equity awards must reflect the current valuation of the company. In the extreme case, the new round renders existing equity worthless for all practical purposes. This is typically referred to as *cramdown* or *washout* financing. By choosing this mechanism, you are not penalizing former employees and common stock-holders. Rather, you are doing what is necessary to keep the team together and make sure the company lives to see better days.

- **Issue a new class of common shares with the purpose of introducing liquidation preferences for current employees (*junior preferred stock*).** This is tricky, and typically would only be done for superstars on the team. The more common instrument is a *carve-out*, which will be discussed in the section "Protecting the Team" in Chapter 12.

- **Negotiate cash bonuses for key employees tied to certain milestones.** Cash bonus provisions can be implemented by amending employment agreements. There is no need to rewrite the company charter or issue stock grants. Cash has the added benefit that it may be the only tangible and reliable means of reward. When common stock is diluted and rendered worthless by the preference table, cash is the much-preferred incentive tool.

Wiping Out Preferences

A late-stage (down-)round provides an opportunity to shed some of the dreadful liquidation preferences awarded to investors in earlier rounds. New investors don't like to honor the benefits and rights of investors who are no longer part of the party—that is, investors who are not contributing new funds. Consistent with a pay-to-play provision in the investor agreement, investors who are no longer *paying* don't get to have anymore say (see Chapter 8). They are stripped of their preferred rights, and their holdings are converted to common stock. Being *converted to common* means that the investor can no longer participate in the investor vote and that the investor's liquidation preferences are being eliminated. Venture funds take great pains at holding large reserves so that they will not find themselves in precisely this unfortunate situation. However, even VCs at times underestimate the future needs of a portfolio company in a difficult market or in an economic downturn. Also, a fund may consciously decide to abandon a portfolio company, even if the fund has the means to continue to support it.

When the pay-to-play mechanism kicks in, the investors do most of the fighting among themselves. They have an interest to keep the preferences in check, so they'll kick out as many members of the *preferred club* as possible. Yet, it is the common stockholders who may end up benefiting most from the reduction in preferences when the company goes through a liquidity event down the road. Beware, though, that sometimes the holders of common stock can get caught in the crossfire, and this dynamic can lead to a cramdown of common stockholders as well.

STRATEGIC INVESTORS

"Once you take money from strategic investors, you will never be able to get money from traditional VC funds again," our CEO and ex-VC used to say. "Strategic investments are a one-way street and a dead end!"

"What's wrong with that?" I countered, thinking that we would never need any more money for ThingMagic anyhow. . . . Only months later, as our once seemingly exorbitant pile of cash was dwindling, my arrogance disappeared about as quickly as the money in the bank.

Strategic majority investors can be a source of funding of last resort when nobody else wants to belly up any money anymore. This is not a situation you want to be in, but it is one that you can address constructively by turning to a corporate sponsor. *Strategic minority investors*, on the other hand, can come in handy when you are trying to embellish a round of funding led by a traditional VC firm.

Strategic Minority Investments: Getting to Know Each Other

The term *strategic investor* is a powerful semantic construct that suggests that the investor, typically a large corporation, can have its cake and eat it too: the strategic investor partakes in the financial return on the investment and enjoys operational benefits for the corporation's operating business. Who wouldn't fall for that proposition?

Big technology companies are bad at innovation, while small companies are good at it, as Clayton Christensen has taught us so eloquently in *The Innovator's Dilemma*.[2] Hence, rather than innovating themselves, large corporations like to stay close to those startups they think might come out with the next big thing.

ThingMagic's A round included a substantial minority investment by Cisco and a Taiwanese contract manufacturer by the name of IAC. Cisco was attracted by the prospect of RFID technology generating lots of network traffic, which in turn would drive the need for more network equipment. IAC was attracted by the prospect of manufacturing the RFID devices ThingMagic was designing.

A minority investment from a strategic player in the industry is a win-win situation in most scenarios. From the point of view of the startup, the benefits are numerous:

- **If you want to buy us, feel free to.** Strategic investors typically require a right of first negotiation (ROFN) in the case that an acquisition offer is extended to the startup. While such a clause looks constraining, it actually is an asset. If the investors are seriously interested in an acquisition and there is another offer on the table, the two parties are going to be competing bidders anyhow. If the strategic investors are not interested in an acquisition, they'll try to maximize the money they get out of the transaction much as the other investors would.

- **Look who believes in us!** Getting a large corporation in your industry to invest means lining up your interests with a strong player in your industry. When you need to convince a potential customer that your little startup is real, the implicit endorsement of the larger industry player can do wonders.

- **Information is provided on a need-to-know basis.** As a minority stockholder, the new strategic investors will not be entitled to board seats. Hence, you are in control as to how much information you are sharing. If you don't want the strategic investors to know about the latest and greatest product development effort, you simply don't tell them.*

- **You changed your mind. That's fine too.** If the strategically motivated investors lose interest in your company, they will quickly expire as active members of your investor group, and you don't have to worry about their opinion anymore.

In summary, strategic minority investors offer a lot of value and no real negatives. When someone offers you an extra million, take it. Just make sure that you are not closing the door on another strategic investor or partner who is more important or is willing to put in more.

With the RFID dream evaporating, neither Cisco nor IAC realized their objectives behind their minority investments in ThingMagic. However, the strategic reasons that had motivated them to invest initially were sound, and they certainly helped ThingMagic financially through a difficult period.

Strategic Majority Investments: Money of Last Resort

At a time when no VC, bank, or angel investor believes in your company anymore, a majority investment by a corporate investor could be your last possible path forward. What kind of strategic majority investors should you be looking for, and what can they do for you?

Potential Acquirers. First on the list of strategic investors are corporations who might consider the acquisition of your startup in the future. If you are honest with yourself, you'd rather have them buy you right then and there, but sometimes you can't have it your way, especially not when times are tough. Also, you can't really tell them directly that you wish they would buy you out. Rather, you beat around the bush

---

* As stockholders, investors do have a right to certain information subject to confidentiality agreements. You should insist on a conflict-of-interest clause in your agreement with the strategic investors.

using lines such as, "We would consider selling if the price were right, but we are not sure you can afford it. Why don't you make a strategic investment instead? That way we'll get to know each other a little bit better, and you keep your options open for the future."*

The potential acquirer-turned-strategic investor will insist on a *right of first negotiation* if and when the startup receives an M&A offer from a third party. The ROFN stipulates that the investor is given the opportunity to make an acquisition offer when another party expresses interest. There is nothing wrong with such a clause, as long as you avoid the scenario in which a third party does all the work on a deal and then is asked to hand the deal over to the investor invoking the ROFN. You have to set the process up in such a way that the ROFN kicks in no later than the time that a term sheet is presented, which is before any third party spends significant resources to validate the deal.†

You can agree to an option to invest more money later, but under no circumstances agree to an option to buy at predefined fixed terms.

Corporate Venture Funds. Venture funds owned and operated by large corporations are usually not mandated to make strategic investments. They are primarily mandated to generate good returns. That said, some large companies use their in-house venture capital group or division to administer any investments they make. In this case, the division or group that decides to make a strategic investment hands the negotiation of the deal and the administration of the portfolio company off to the venture division.

Some corporate venture funds behave very much like regular VC firms. Others have more of a corporate culture. The differences are partially rooted in the investment charter and in the way executives are managed and compensated. Partners in VC firms are compensated

---

* What you really mean is you would always talk to them about an acquisition, you are desperate and could be bought for very little cash right now, and you will always sell to the highest bidders.

† It is less common to offer investors the *right of first refusal* (ROFR), which means the company has to notify the investor about a bid it received and then allow the investor to match it to acquire the company.

via limited partnership agreements that stipulate management fees and carried interest. Corporate VC funds tend to compensate principals with salaries and bonus programs, promoting a more balanced set of objectives relative to portfolio companies. The financing of a financially underperforming, technically strong venture at a bargain price is out of the question for a VC fund; it may, however, be an OK investment decision for a corporate fund.

Before you try to raise money from a corporate fund, find out how the company is organized and how it selects and values portfolio companies. The more you understand the particular culture, the better you can adjust your pitch and sell the corporate VCs on your venture.

Customers. For a variety of very good reasons, few startups turn to their customers to seek funding. Mixing a customer relationship with an investor relationship is tricky and usually the very last resort. However, if you are really hurting for money and are faced with the prospect of going out of business, your key customers may actually be the stakeholders that get hit the hardest. If they are using your product in a critical part of their business, losing your supply and support could be extremely tedious.

As you are trying to hit your customers up for money, very carefully suggest to them that an investment would align your interests, or, if things are obviously bad, that you need their help to keep helping them ("Help us help you."). Almost certainly you will have to agree to special terms, including the following:

- **If the patient has multiple demanding customers: offer preferred treatment on product shipments and support services.** The investing customer will be concerned about ensuring the supply of product and support. In exchange for funding, you might have to agree to serve them first before shipping to their competition.

- **If the patient improves: offer the right of first negotiation on an acquisition offer.** No customer-turned-investor likes to find out later that the money they invested was used to sell the startup to a competitor.

- **If the patient gets worse: offer collateral in the form of intellectual property.** The investing customer will want to be prepared for the distinct possibility that things get even worse. While a conventional investor would ask for financial collateral, an investing customer would be most interested in access and rights to the technology and products.

- **If the patient dies: offer escrow arrangements for the technology.** The usage right to certain intellectual property is important, but so is physical access to the information. An escrow agreement calls out the rules by which a bankruptcy triggers the release of critical product and design information. When nobody picks up the phone at the startup anymore, the customers can ask the escrow agency for access to the information.

WHEN THINGS GET REALLY DESPERATE

Lesson 1: Don't get depressed because you need to raise a down-round. Console yourself that you must have done an excellent job marketing your company's optimistic valuation in the first place. You are probably better off than you would be if the last round hadn't happened at the valuation at which it did.

Lesson 2: Try to work out a deal with your existing investor group first. You have better leverage with investors who are already committed to you, and they are more likely to invest on acceptable terms.

Lesson 3: A depressed financing round with new investors is not going to be pretty whichever way you put it. As you negotiate with the new investors, focus on the terms and implications of the round for your employees. The new investors should want the remaining employees to be happy and to stay around.

Lesson 4: If you really can't find money anywhere else, turn to potential acquirers, strategic partners, and even key customers. All of these guys want to see you succeed.

CHAPTER 10

Right-Sizing

Not that I lov'd Caesar less, but that
I lov'd Rome more.
        —WILLIAM SHAKESPEARE (1564–1616)
        Julius Caesar

Startups are never quite the size they should be. First they are too small to handle the opportunities in front of them, so they hire. Then they become too large to sustain themselves, so they need to shrink. While large corporations can survive with a bloated payroll for some time, the margin of acceptable error in a small technology company is tiny. If you don't reduce your staff or you don't do it quickly enough, chances are you will lose the entire company, not just a subset of your employees.

Each time we were facing drastic reductions in staff at Thing-Magic, we pushed out the inevitable for months. Each time, the reduction would happen after all. As hard as it is, it is your responsibility to reduce personnel when needed. Don't get emotional about it! Remember that your loyalty is to the enterprise as a whole, not any one individual. Your primary job is to protect the future of the venture and to preserve shareholder value.

FIRING FOR NONPERFORMANCE

Being your own boss, you would think you can easily get rid of an underperforming employee. Quite to the contrary, *termination for nonperformance* is an often lengthy and tedious process that typically leaves the employee as well as her manager exhausted. The company should follow due process so as to protect itself against possible legal action.

In essence, you need to give any employee a second chance and help her get back in good standing. The process implements what is called the *performance improvement plan* (PIP). It is not legally required but good practice if you don't want to get sued later:

- To initiate the PIP, invite the employee to a meeting during which you present a detailed, written statement of the observed deficiencies. You also provide a written description outlining what the employee will have to do to cure the situation and how her progress is going to be measured. The duration of the PIP should be such that the employee can reasonably be expected to deliver on the outlined tasks typically in 4 to 12 weeks' time.

- You then meet with the employee regularly to monitor and discuss progress toward the deliverables. Each meeting should result in minutes that state clearly if progress is being made or not.

- At the end of the process, the manager decides whether the deliverables meet the expectations established in the PIP. If the employee has improved, she can stay; otherwise, the manager can terminate her employment.

This process does not eliminate the risk of a legal claim, but it greatly reduces it. It is important that all communication be backed up by documents. Verbal communication with the employee alone doesn't do the trick.

Be careful in performance reviews preceding a PIP. If you just wrote a modestly favorable review of someone, don't count on firing her for nonperformance anytime soon. It is just going to be impossible to explain how the individual's performance deteriorated so quickly.

At the time that a reasonable manager decides to put an employee on a PIP, she likely has already tried everything to improve the situation. Hence, the PIP just formally demonstrates that the company has tried everything to keep the employee. In my experience, very few PIPs result in the employee staying on as a productive member of the team.

REDUCTION IN FORCE

"Well, anybody who ever built an empire or changed the world sat where you are right now. And it's because they sat there that they were able to do it. That's the truth." Ryan Bingham, George Clooney's character in *Up in the Air* (2009), is in the business of firing people. He travels around the country to help employers lay off staff.

Having to tell employees that they are no longer needed or affordable is just about the most unpleasant task in a startup manager's life. I would have liked to have handed off the task to Mr. Bingham, but that would have been in bad taste, and we didn't have the money to hire a consultant.

Here you are, management or the board of directors has decided to send a good portion of the staff home. How do you go about this most unfortunate of all startup events?

How Many?

Once the decision has been made to implement a reduction in force (RIF), you first need to decide how many to lay off. The board probably approved a financial plan that suggests a certain number and type of employee. Your natural reaction is to fight for every single individual; you try to hold on to as many people as possible. Instead, you should consider carefully if a larger layoff might be better.

As you are *sizing* a RIF, make it your most important objective that there not be any additional layoffs in the foreseeable future. You are better off creating some financial margin for the company, so you don't have to do it all over again soon thereafter. As you go through the painful process of laying off employees, also see to it that you are left with truly exceptional people. If there are B players on your staff, now is the time for them to move on to other opportunities. You are better off hiring exceptional new candidates when business picks up rather than burdening the team with employees who are not pulling their weight in difficult times.

Is There Ever a Good Time?

There is much debate about which day of the week is most suited for a RIF. Monday, say some, is best because it allows the affected employees to start looking for a new job on Tuesday. That keeps them busy, and they don't develop unfortunate ideas. At the same time, the remaining employees come back on Tuesday and realize that the place is still up and running. *Friday* is best, others say, because it allows affected employees to spend the weekend with their families before reality kicks in the following Monday. Unfortunately, RIFs are unpleasant whichever day of the week they happen.

Similar arguments are made relative to national holidays or individual vacation time. Some say you should always have affected employees enjoy their last holiday, after which you can break the news to them. Others recommend to lay off people just before the scheduled holiday so that they can get on with their lives and prepare for the future. The latter approach seems more sensitive to the needs of the employees. Wouldn't you want to know what's coming before you spend money on a family vacation? The financial implications for the company are the same, by the way. In either case, the employees get paid for any accrued and unused vacation time.

Severance Packages and Severance Agreements

Even if you don't have much cash for severance pay, make sure you put enough benefits together so that you can call it a "severance package."

Try to include at least a few days of salary payments, and sweeten the deal with some placement services. In addition to a severance payment, you can offer to support the employee's claim for unemployment benefits. By law (Consolidated Omnibus Budget Reconciliation Act of 1985 [COBRA]), you will also have to administer the health insurance benefits for parting employees for a year and a half following the departure of the employee.

In exchange for the severance package, the departing employee is expected to sign the *severance agreement,* which includes terms to protect the company. Most important, the severance agreement reconfirms any noncompete and confidentiality arrangements, and it includes a general release of the company.

By law, employees must be given at least a few days to decide whether to sign the severance agreement. However, most employees need the money and clarity rather quickly. The signed documents tend to show up way before the consideration period ends.

If the company employs 20 or more and if the group of laid-off employees includes anyone over the age of 40, the whole process gets more complicated. The *Older Worker Benefits Protection Act* (OWBPA) stipulates that over-40-year-olds have up to 45 days to decide on signing a severance agreement, and they must be given additional time to revoke their decision.* In addition, the company must notify the older employees about specifics of the layoff, including the job titles and ages of all the employees laid off, and the job titles and ages of employees not laid off.†

Departure Meetings

The departing employees are given the news in individual meetings, which should take no longer than 30 minutes, but could be significantly shorter. It is best if word of the layoffs doesn't get out before

---

* This applies only if the employees waive their rights under the Age Discrimination in Employment Act of 1967 (ADEA).

† Consult an employment lawyer to understand the full requirements of OWBPA and which specific rules are applicable in your situation.

the last affected employee has started her meeting. This means you need to schedule multiple parallel *tracks* of interviews. By the time the company knows from the first *shift of interviews* what's going on, the *second shift* is having their meetings.

The meetings should be attended by at least two people representing management: typically the immediate manager of the employee and a representative of HR. During the meetings, it is best to simply present the facts: (a) "You and others are being laid off for the following reasons"; (b) "These are your rights, and this is the severance package that is being offered"; (c) "Here is what will happen in the coming days." There is no room for discussion or negotiation. Empathy is okay, but any advice you might give should focus on the professional and personal future of the employee.

Packing Boxes

While the individuals are at their exit interviews, their e-mail and network access should be terminated. Immediately following the interviews, the individuals are asked to surrender their computers, hand in their keys, pack their personal belongings, and leave the office.

I was quite outraged when this cold and cruel procedure was explained to me for the first time. But once you have implemented the process a few times, you begin to appreciate the rationale behind it. You just don't want a disgruntled employee to steal vital IP, publish source code on the Internet, talk others into quitting, or make a scene in front of everybody. None of this usually happens, but when it does, it can have serious repercussions.

Company Meetings: Taking Care of the Survivors

The company meeting immediately following the exit interviews is the most important event on the dreaded layoff day. When all the departing employees have left, you need to convince everybody else that (a) management is not a bunch of jerks who are quick to fire esteemed colleagues and friends and that (b) the company is going to be all right and successful in spite of or because of the RIF. Be prepared to talk about the strategy of the company going forward and how the change in direction required a change in personnel. Make it clear that the company needs everybody left to be fully committed and working hard. Subtly suggest that the employees who did not get laid off are the high performers and are the foundation for the future success of the company. However, do not say anything that is inconsistent with the message provided to the departing employees, since such a statement could be used in a discrimination claim.

Most important, convince your employees that the RIF was a one-time event that will not be repeated. If you fail to get that key message across, your employees will start looking for other employment the very same evening. Naturally your best people will be successful in finding new jobs. If you are planning on hiring different skill sets in the near future, explain what those positions are going to be.

When the layoff happens, the writing has usually been on the wall. The rank-and-file employees have had some idea that the company wasn't doing as well as it should to pay for everybody around. So everybody is expecting a reduction in staff of some sort, and, of course, most everyone is fearful of being axed. When it happens, you might think that the survivors will be upset about their colleagues leaving. In reality, those who stay are likely relieved that they can go home to their families that night without having to worry about their paycheck for the time being.

Serve food at the meeting to convey the notion that the company still has money in the bank and will continue to pay its payroll! Don't go overboard with the catering though. Sushi might be considered in bad taste.

Public Relations: How to Turn Bad News into Good News!

A layoff is not something you typically advertise. It is perceived as a sign of weak performance (usually true) and an indicator of future trouble (not necessarily true). However, even without a press release, the news of your RIF will spread like wildfire in the industry. So whether you like it or not, your major customers and partners will know about what happened very quickly.

A good way of counteracting the negative rumor mill is to issue a positive press release about an unrelated issue—for example, the release of a long-awaited product, a major partnership, or a change in direction. Readers of the press release will understand the news about the RIF in the context of your company actively implementing a new strategy. The RIF will be seen as a proactive event to achieve a specific goal, rather than as a reactive measure to deal with financial difficulties.*

The stock price of public companies routinely goes up when companies announce major layoffs. The rationale behind this phenomenon is that the RIF is expected to have a positive impact on the financial performance of the company. If you are looking to raise additional funds for your startup, a layoff may benefit you in a similar fashion. Investors prefer to invest in companies that have their costs under control.

What to Say

To be told that you are no longer needed is devastating. Be prepared that nothing you say can ultimately change the fact that you just pushed someone off a metaphorical cliff. As hard as it is for the employee, it is an exhausting, unpleasant, and traumatizing task for the manager as well. Certain statements and phrases can help you get through the event with fewer emotional bruises (see Mr. Bingham above). Think about how you want to open the conversation ahead of time.

Sometimes, you may get help from the employee:

> Employee [upon entering a room occupied by her manager and an HR person]: "Is this what I think it is?"

> Manager [relieved that he doesn't have to say it out loud]: "Unfortunately, yes. We are very sorry."

---

* Again, make sure not to publish anything that is inconsistent with the reasons given to the parting employees.

Hopefully, you are lucky enough to get through the event without incident. But in case things don't go as planned, you might want to know what to do if any of the following occurs:

- **Someone starts crying during the exit meeting.** Emotional outbursts happen and are understandable. Employees seem to take the news better if they are told that they are not the only ones who are affected. Emphasize the message that a whole group of people needed to go.

- **Someone threatens legal action.** Threats are made easily in the heat of the moment, but they are usually later abandoned. Respect the rules, follow the process, and do your best to keep calm.

- **Someone instigates an uprising among remaining employees.** A departing employee who immediately starts a campaign to convince colleagues of the unworthiness of the management team or company should be ushered out. Someone who just wants to say goodbye or is in need of friendly consolation should be given some slack.

- **Someone gets physical in the office.** I have never experienced such an unfortunate incident, but I admit that I hired a security guard for our very first employee termination. I decided it was better to be safe than sorry, and I wanted to be prepared if something went awry. We hid a scary-looking guard in an empty office. Fortunately, he didn't have to come out during the event.

SHOWING EXECUTIVES THE DOOR

In a short period of a few years, ThingMagic hired and fired five senior sales executives. The executives were not able to bring in sufficient revenue for the company to thrive, but all of them cost us dearly, both in money and in time. We will never know for sure if they failed because of who they were or because of the company and the market we worked in.

As with rank-and-file employees, we waited far too long to let these and other executives go. We liked them as people, and they did try to do a good job. The trouble is that it takes quite a while to determine that an executive is not working out.

Sales executives are easier to evaluate than others because their performance comes down to a few key metrics. However, you have to cut them some slack early on because you can't expect them to produce revenue on their first day. It also seems unfair to fire them when the competition isn't selling anything either. On the other hand, if she isn't selling, it doesn't really matter whose fault it is. You had better let her go and save the money to live another day.

When letting executives go, sometimes the process will be no different from the description above. However, if at all possible, try to part ways *by mutual agreement* and allow the executive to manage the message on her way out. While it is preferable and a lot less stressful if the executive is actually in agreement, establish some notion of *an amicable separation* even if she yells as she's going out the door. As you announce the departure, *customize* your message depending on whom you are talking to:*

- **Addressing your employees.** "Effective today, John is no longer with the company. We jointly came to the conclusion that the company needed a different kind of talent in the position. John continues to believe in what we are doing and sends you all a warm farewell."

- **Addressing the board.** After having presented John as god-like a couple of years earlier when you had to justify his outrageous signing bonus and headhunter fee: "John wasn't right for the job. We lost out on a number of great opportunities because of his lack of presence and skill. It was time for a change. We have secured John's full cooperation during the transition."

- **Addressing John's reports.** "As you know, John was falling short in a few areas, so we mutually decided to part ways. We have

---

* Make sure to stay consistent with any nondisparagement clause you might have signed.

started to look for an immediate replacement. In the meantime, we count on your hard work to make sure we are not falling any further behind."

- **Addressing customers.** "John has left the company to pursue other opportunities. Your day-to-day needs will be taken care of by your new account manager. We assure you that nothing will change in the way we are doing business with any of our customers."

Try to let executives go one by one, even if the departures are related. It is easy to explain one executive's departure with her lackluster performance. When more than one executive leaves at the same time, it looks as if the company has hit hard times. Even worse, employees and outsiders may conclude that the management team has lost faith in the company's future and is jumping ship.

Termination should not come as a total surprise to the individual. Approach the topic slowly to gauge if the executive has herself come to the conclusion that the employment relationship has not been working out and that there would be a change. Salespeople tend to see the writing on the wall. The numbers just don't lie. Executives in other departments are often less willing to accept that their contributions are not satisfactory or that their skill sets are misaligned with the needs of the company. It can take a few meetings to make the point and get their buy-in for the change.

Once you have a basic understanding that the executive's employment is coming to an end, work with her to define a transition plan.

Standard Terminations

In a standard termination, you agree on a last day of employment, after which day the executive's relationship with the company is ended for all purposes. You present the executive a severance agreement, including severance pay. The actual amount depends on how dire the company's financial situation is and how similar situations have been handled previously. Try to avoid the establishment of past practice, but if you have always paid three months of severance, don't change that policy suddenly or you may end up in court.

In exchange for the severance payment, the executive agrees to a subset or all of the following terms:

- A reconfirmation of all confidentiality and noncompete provisions as stated in the executive's employment agreement with the company.

- Possibly additional noncompete provisions relative to a particular competitor, product, or business model. Within the period of the executive's employment, the market and the competitive landscape likely have changed. The severance agreement offers the opportunity to protect the company accordingly.

- A nonsolicitation clause. The executive should not be allowed to directly or indirectly hire your employees upon leaving.

- A nondisparagement clause. This applies to both the executive and the company. Neither should speak badly about the other. You can be detailed in specifying exactly what to write in a reference request should the company get one.

- A general release of the company with respect to the termination of employment and any other matter.

Transition Plans Without Severance Pay

In this case, the executive agrees to a similar severance agreement. However, instead of severance pay, she will continue to work for the company for a well-defined period of time, either as a full-time employee or under a consulting arrangement.

During this transition period, you should change the responsibilities of the soon-to-be-gone executive. Assign a new manager to her reports. Pair her up with another sales manager when talking to customers. Monitor her business e-mail during the transition period (with her consent), so as to be able to pick up any customer leads and other communication.

Assign the executive to a project that is somewhat unrelated to your core business. For example, have her research a new market segment or a new product. In the worst case, the executive does not produce. In the best case, you will get a business or marketing plan for a new initiative at the end of the transition period. It will cost you no more than you would have had to pay to protect the company from legal action against unlawful termination.

IT'S NOT YOUR FAULT

Lesson 1: Managing the ups and downs of a startup inevitably includes terminating employees. You cannot afford to be emotional about it. You need to implement staff reductions when they are necessary to protect the venture as a whole.

Lesson 2: When terminating an employee for performance reasons, you need to be able to demonstrate that you gave the employee adequate notice of her deficiencies and an opportunity to improve. Proper documentation is critical.

Lesson 3: When you lay off people, focus your attention on the employees who are staying, especially the high performers. You have to make sure they remain comfortable with the future of the company so they will not start looking for another job.

Lesson 4: Try to sever executives amicably by offering them a *transition plan*. It will save you money and emotional energy, and it more likely leaves relationships and friendships intact.

PART III

EXIT:
SELLING YOUR BABY

CHAPTER 11

Startup Dynamics in Crisis

> *Life is not a matter of holding good cards, but about*
> *playing a poor hand well.*
> —ROBERT LOUIS STEVENSON (1850–1894)

In a startup, it can be difficult to distinguish times of actual crisis from business as usual. After a few years, it mostly feels like one long crisis—never boring, but certainly stressful. Eventually, you may find yourself so sick and tired of the whole thing that you really wish you didn't have to go back to work the following morning. Yet, you have invested so much of your own blood and time that you really can't walk away. You might as well try to get something in return for your sacrifice!

TIRED FOUNDERS

Entrepreneurs make for a restless bunch. When they start their ventures, they pour all their personal energy into the enterprise—and then some. They work passionately and forget about everything else life has to offer. Yet, when the initial enthusiasm meets with the difficulty of growing the venture to the next level, company founders are easily overwhelmed with disillusionment and fatigue.

The long hours, the traveling, and the uncertainty are exhausting, and all of it makes you want to give up. The fact that you have not been able to get to a quick exit or to extraordinary growth makes you feel like you are missing a bigger opportunity somewhere else. These two emotions reinforce each other. Why should I work so hard if the financial reward is not coming? Why should I work here when my buddies are running successful businesses elsewhere?

When I reached low points of disappointment and frustration, I reminded myself that I had the most desirable job in the world. It wasn't just an empty psychological trick. I really did not envy any one of my friends and acquaintances in corporate jobs or academic positions. In conclusion, I had to consider myself happy (or at least happy by comparison), and I continued working for the company I cofounded. Unfortunately, it is not always that straightforward (it wasn't for my cofounders), and the frustration of a founder can quickly get out of hand.

When a founder gets tired, bored, or unmotivated, and he is ready to move on, it may be next to impossible to convince him to stay. Financial arguments usually won't cut it. It is likely the founder already owns a lot of (possibly utterly worthless) stock. A few additional options won't really make a difference. Additional cash payments are usually helpful, but if the company is not doing well, cash is difficult to come by. Also, investors don't like the idea of throwing additional compensation at an unmotivated member of the team.

Executives who are no longer pulling their weight or who turn into complainers can be replaced. With founders it is not as easy. The energy of the founding team or the larger-than-life personality of an individual founder counts among the key assets of a small enterprise. When the founding group is no longer there or not working as hard as they used to, the livelihood of the small company is threatened: investors are unwilling to provide more funds; employees lose their role models; and customers lose confidence in the long-term viability of the startup as a supplier.

In a company with multiple cofounders, the dynamics are particularly complicated. At ThingMagic we had to manage through the departure of three cofounders before the company was sold. Each of the departures happened for different reasons. Yet all three were highly dramatic and made us wonder whether the company would survive. The last departure took place a couple of weeks before the company was acquired. We were terribly scared that the event would set off an unfortunate chain reaction of employee unhappiness, more departures, and possibly the collapse of the M&A deal.

When a founder or key contributor makes it clear that his time with the company is coming to an end, the only viable path forward may seem to sell the venture. If the founder is no longer committing his future to the venture, why not call it quits and get as much money for the company as possible? Unfortunately, this is where the trouble starts, not ends. Any reasonably experienced acquirer will insist that the management team, including the founders, stay on and be involved post-acquisition. Unfortunately, the unique importance of the founding team extends through an acquisition and beyond.

In the eyes of an acquirer, the departure of a founder reduces the value of the company significantly. This dependence puts a lot of pressure on the core team to stay put, whether or not they would rather do something else with their lives. In response to some founders not playing along during the saga leading up to ThingMagic's acquisition, some of our investors came up with rather questionable arguments to secure cooperation. One stakeholder made the interesting—but utterly false—claim that the venture capital industry relies on the implicit, moral obligation of startup management teams to continue working for an acquiring company.

If a founder cannot be persuaded to stay on and see an acquisition through to success, the negotiation with the acquirer is certainly made more difficult. But, if carefully handled, the situation can be salvaged. What is to be done to counter the appearance of weakness due to the departure of a founder during an M&A transaction?

You must find a credible reason why the founder is leaving and why it is for the best of everybody involved. Most likely the founder

THE TECH ENTREPRENEUR'S SURVIVAL GUIDE

is upset, doesn't believe in the business and its financial viability, and hates the thought of spending any more time with the products he co-invented. The translation of this reality to the potential acquirer should read like: "The founder is no longer personally interested in the venture's technology and wants to move on. He is looking to do something in a completely different industry." Make sure you stress the part about *different industry*. The prospect that the departing founder might set up a competitive business would certainly be a deal killer.

Acquirers understand that founders are motivated by lifestyle considerations and personal preferences that may have nothing to do with the venture or the acquiring company. The financial reward and retention bonuses are often just not enough to compensate for the personal sacrifice of staying on. Also, as bad as it is to have a key employee depart, it is not as bad as having a key employee hang around unhappy and unproductive and prevent the hiring of a suitable replacement.

Finally, play the *it-is-better-this-way* card. Argue that the company needs different talent to take the venture forward, now that it is part of a larger entity. Seasoned managers should take over and replace the playful founder!

IMPATIENT INVESTORS

Investors like to see returns sooner rather than later, and they like to move on with their investment activities as quickly as possible. A typical fund is expected to liquidate within 5 to 10 years. Way before liquidation, the VC team starts raising and investing the next fund.

Hence, simple practical matters alone bias VCs toward liquidating portfolio companies when there is an opportunity to do so. Beyond the normal course of VC investing, a myriad of other reasons can cause investors to push for an exit, whether it is prudent strategy for the startup or not:

- **We would like a bigger return, but time is up.** Equity investors all want to maximize their return on investment. However, they differ in their expectations of timing and level of risk taking. An angel investor may very well be content to contribute to the local startup scene and get some of his money back no matter how much. A high-profile VC firm, on the other hand, might consider an exit only if it generates 10X in return or more. Depending on the expectations among your investors, they will encourage you to exit or not.

- **Oops, we are out of money.** Professional investment funds set money aside so as to be able to participate in follow-up financing rounds of portfolio companies. If they don't have any more such funds available, they lose all leverage relative to their fellow investors. To avoid this course of action, the investor may push for an exit before any new money is raised.

- **Who would have thought that our balance sheet could ever look this awful!** Private equity funds hit hard times occasionally, prompting them to push their portfolio companies into liquidation. If the fund has leverage among the investor group, the situation can paralyze the portfolio company and force a sellout.

- **We like to pretend it is our money, but it actually is not.** Venture capital firms are ultimately money managers for limited partners. While the limited partners typically do not have any direct decision power regarding individual investments, the opinion of the investors certainly matters. A general partner may decide to force portfolio companies into selling out because an influential limited partner wants to withdraw his funds.

Going Cash Flow Positive

The investor's leverage diminishes significantly if the company is operating at breakeven. A venture without need of additional operating funds can almost certainly defend itself against investors wanting out. Of course, generating positive cash flow is easier said than done.

Realistically, getting to a positive bottom line means reducing expenses quickly. Since eliminating the "L" in P&L is hard without reducing staff, a quick reduction in force may be the only strategy to ensure the independence and livelihood of the remaining entity. Staff reductions typically need board approval, but they may not be subject to the investor block.

Restart/Restructuring: Let's Try This Again with a Different Investor Team

Rather than letting the company run into bankruptcy or foreclosure, the management team and investors should consider restructuring the company proactively to give it another chance. If the company is at least marginally profitable, this can mean that the investors give up their preferred rights in exchange for common stock. With preferences eliminated, the management team is more likely to keep the employees motivated, carry the business forward, and attempt an exit in the future. If things do work out, everybody benefits, including the investors.

If additional capital is needed to carry on, new investors who are willing to fund the operation going forward have to be found. The old investors need to be prepared to endure significant dilution to allow the new investors to take over. This option also provides the opportunity to free the venture of common stockholders who have left the company and are no longer helping to make it successful. It's a fresh start for everybody involved.

Bankruptcy: We May Buy You Out, but Only if the Price Is Right (Rock-Bottom That Is)

In a bankruptcy filing senior creditors are satisfied first, most importantly the banks. The remaining assets are then distributed among the preferred stockholders in order of seniority of their stock. Keeping control of the venture in the hands of the founders and executives following a bankruptcy is hard and requires the cooperation of the existing creditors.

If the management team wants to continue the operation of the company and stay in charge, bankruptcy is a good time to orchestrate a management buyout. A new investor group has the opportunity to buy into the company at a bargain price, while the existing investors need to be prepared to get back pennies on the dollar. The management team is well advised to understand the financial prowess of the new investor group. You don't want to find yourself in an unfortunate and hostile situation again.

A Controlled Bank Foreclosure: Measure of Last Resort

A controlled bank foreclosure is a possible, yet tricky, strategy for defending yourself against investors who are no longer cooperating. With a bank loan outstanding, the creditor institution can take control of the company if the financial situation of the venture deteriorates. The lender imposes financial covenants to monitor the health of the company and contain the likelihood of a default (see the section "Banks and Loans" in Chapter 5). When the metrics are not met, a bank has the right to foreclose the company, take ownership of the assets, and sell them to anyone it chooses.

In reality, however, a bank does not like to foreclose on the first violation of such covenants. Instead, the bank prefers to negotiate with the company to restructure the loan. When things are really dicey, the bank seeks the cooperation of the team to jointly find a solution that allows them to recover their principal and interest. This can mean a sale of the assets and the transition of certain employees to the new owner. It can also mean a management buyout, in which the management team takes the lead on a restructuring effort, including the search for a new backer who covers the money owed to the bank.

In either case, the foreclosure action wipes the capitalization table of the company clean. Existing stockholders lose their ownership interest, and the bank gets the cash proceeds from the restructuring effort. Don't forget to re-incentivize your employees with new equity grants. Now is a good time to do so because the value of the entity is once again very low.

Warning: playing the foreclosure game in many ways is a last resort, comes with significant risk, and may backfire. Don't do this unless you absolutely have to, and make sure to have good legal representation!

Fighting Back with Leverage

If investors are insisting on a sale of the company and none of the strategies outlined so far are viable, management still has degrees of freedom to minimize the negative impact of an unwanted exit. Management and employees have leverage because small technology businesses are impossible to sell if their expert staff leaves (see also the section "Protecting the Team" in Chapter 12).

Investors have no interest in anything that happens after the acquisition. The acquirer, on the other hand, is interested only in what happens after the transaction. Your new bosses will try to tie the compensation of the executives and employees to future performance. You need to be vigilant that such incentives are structured fairly. While it is perfectly reasonable to set up retention and performance incentives for employees, it is not reasonable to put all of the employees' compensation at risk and have the investors walk away with all their cash at closing.

If the acquisition is structured as a stock deal, make sure that sale restrictions on the stock are consistent among the different recipients. Unless the stock is awarded as a retention incentive, lockout provisions and other restrictions should be similar for investors, common stockholders, and employees.

WHEN TO SELL?

The opportunities to sell a high-tech venture on decent terms are few and far between, an insight expressed in the old adage that "companies are bought, not sold." When the startup is doing badly, selling is obviously difficult. When it is doing well, opportunities to cash out are easily overlooked given the general excitement and the prospect of multifold gains down the road. And yet the next opportunity to sell may be years away or never materialize at all. Hence, when you do come across an opportunity to exit or advertise your company for sale, take advantage of the precious moment as best you can.

In hindsight, I know the exact time when we should have sold ThingMagic. Our biggest competitor had just been acquired for an extraordinary sum, RFID technology was the next big thing, and every large technology company wanted in. We had previously taken no outside capital, and founders and employees would have pocketed every single dollar from an M&A transaction. Alas, we felt so bullish about our prospects that we did not even consider cashing in. Instead, we heavily diluted our ownership, and then we watched as our industry entered a period of rapid and steady decline. What is it they say about hindsight?

The Untested Hypothesis

As you try to determine if it is a good time to sell your company, use the concept of the *untested hypothesis*.[1] Each time the young venture reaches a decision point or makes a bet on the future, the company poses an implicit hypothesis that its strategy and investments are sound and will pay off. As time passes, the hypothesis is put to the test, and many times it is proven wrong: the market did not develop as predicted; the newly developed product is losing to the competition; the technology is taking longer to be ready for prime time.

It is easiest to sell a company when a major commercial hypothesis has just been validated (case A); it is harder to sell against an untested hypothesis (case B); and it is virtually impossible to sell a company when a major hypothesis has just been proven to be wrong (case C).

Try explaining to a potential acquirer that you were just proven wrong (case C) but that you are nevertheless bullish about the future.

Suppose your product flopped recently, and you have to make the case that the next one will be a big hit. Difficult argument, no? On the other hand, raving to an acquirer about the imminent release of your new product (case B) comes naturally to most startup executives.

Internet companies in the late 1990s posed one of the biggest startup hypotheses ever: hundreds of companies claimed that the number of users and page clicks would eventually translate into revenue and profits. While the hypothesis proved true for a select few (Google, Yahoo!, Amazon), for most dot-coms it turned out to be utterly wrong.

As you are facing a major untested hypothesis in your startup's life cycle, take the time to think about the upcoming validation phase. Consider the realistic prospects of your company and whether it may be better to capitalize on your sweat equity right now. Give your usual optimism a break, and be honest with yourself. You know your industry and your company better than anyone. Use your insights to your advantage before your hypothesis is proven wrong!

Selling Prefunding

Self-funded and bootstrapped high-tech ventures reach a point where the lack of capital becomes constraining. They are tempted to accept outside capital as a means to fund significant product development, marketing campaigns, or sales efforts. After the financing, the return of capital to the investors dominates the decision making, and common stockholders become secondary beneficiaries to any financial upside. If, however, management were to sell the venture instead of raising capital, all the proceeds would go to common stockholders.

A simple example and calculation shows how compelling an exit prior to funding can be. Let's assume a bootstrapped company raises the equivalent of its pre-money valuation—that is, the company's valuation doubles in the funding round, and the ownership of the common stockholders is diluted by 50 percent. Let's further assume that the investors are granted a 1X participating preferred liquidation preference (see also the section "Liquidation Preferences" in Chapter 8). If the company sells after funding, the sales price would have to be three times its prefunding value in order for the common stockholders to take home the same return. Needless to say, increasing the enterprise value by a factor of three is not an easy task. It can take a long, long time to get there, and many of us venture-backed entrepreneurs never make it there at all.

Selling Preproduct

Before you start the development of a new product, the untested hypothesis suggests that the product will work out technically, that it will hit the market on the intended schedule, and that it will be commercially successful. Any one of these assumptions is risky; together they are almost laughably overambitious.

Lots can go wrong in product development. It's expensive; it always takes longer than you think; and predicting what the customers want is virtually impossible. If it were easy, everybody would be developing high-tech products. After all, it is very lucrative when it does work. Technically complex products are almost always never developed in the time frame anticipated. The only way to stay on schedule is to build a significant margin into the schedule in the first place. While large companies tend to do that after having fallen on their faces too many times, startups notoriously do not.

Once you spend a ton of money to do all of the above and then find out that the product does NOT work, you have a real problem. When large companies produce a flop, they get a second chance. Startups, on the other hand, rarely have enough funding to try again. If you are in a position to control an innovative and promising technology, think of a potential buyer who might invest in the product development process instead. Letting an acquirer make that massive bet may be the better option.

There is yet another disadvantage to product development in contrast to the development of a broad and horizontal technology. If you start with a technology, and then develop it into a specific product or service, you inevitably reduce the potential buyer pool for the technology itself. Once you apply your IP in one application area, few buyers in another area will be interested . . . unless, that is, your product didn't come through as hoped, and your startup can be purchased at a bargain price!

Selling Prerevenue

There is an element of *if we build it, they will come* to the commercial release of any technically advanced product. Nobody is able to predict with accuracy how a new and innovative product will be received by the marketplace. Methodical and rigorous product marketing helps limit the risk, but it cannot eliminate the large margin of error on predictions of commercial success.

It is also an unfortunate reality of commercialization that the biggest expenses and efforts happen after the technical development is complete. The engineering and design work are only the tip of the iceberg of the overall market introduction process, especially for hardware products. In order to sell a product, you have to do design-for-manufacturing, find a manufacturer, develop distribution partners, write documentation and training collateral, make sure your product is certified and consistent with regulations, observe export restrictions, apply specialized accounting rules for your business, and so on. You will have to take care of all these tasks to find out for sure if your product is as popular as projected—or not.

It is far cheaper and also a lot more fun to develop an exciting prototype and *sell the concept* to the potential acquirers of your company. A prototype can be fanciful or basic, complex or simple, refined or rough. As long as the device or implementation conveys the essence of the future product, the prototype has fulfilled its purpose.

Most creative technologists derive a great deal of satisfaction from demonstrating novel capabilities. Few technologists get excited about tedious product development and marketing. Before you embark on the adventure of commercialization, do some soul searching within yourself and your company to determine whether you really have a product release in you. The potential for disappointment and frustration is unfortunately high. It may just make more sense to get some money from the sale of your prerevenue venture and then move on to the next exciting startup.

Selling Preprofitability

Suppose you have done everything to commercialize your product. Sales are beginning to come in. Revenue is increasing quarter over quarter, suggesting a very bright future.

Your next challenge is to boost revenue to the level where your gross margin covers your monthly expenses. On paper it looks as easy as two intersecting lines. In reality, however, it is a lot harder than that. Countless startups before yours have failed to reach breakeven for a long list of reasons: change in market needs, competitive pressures, delayed product launches, low gross margins, commoditization, and insufficient channel development. The risk that you will be among the many who keep losing money forever is high. The likelihood that you will be among the lucky few who make a profit in the near future is low.

Consider selling the venture when profitability looks plausible and achievable. The window of opportunity to sell before reaching breakeven is short. Wait for the moment when sales are on an upward trend, however small the slope of that trend may be. Once you have demonstrable growth, your "profitability story" is based on simple arithmetic. As soon as you hit stagnation or decline, your pitch to an acquirer will sound like a fairytale with little credibility.

At ThingMagic we failed to exit when the retail supply chain industry made all of us believe that its hunger for RFID technology was insatiable. Revenue was increasing rapidly—so much so that we thought the boom would last forever. What a missed opportunity! When we finally got around to selling the company, we were lucky to benefit from a couple of quarters of modest growth and even a little bit of retail-related media hype. The retail industry had kicked off the second—albeit small—RFID boom.

The Hard/Easy Quadrant

In graduate school, we used to classify technology and demos according to the following matrix:

Hard/Easy Matrix Applied
to Technology Demos

	Looks Easy	Looks Hard
Is Easy		Good demo: the most bang for the buck!
Is Hard		

Since we never had much time to prepare, the much preferred option for a successful demo was the looks hard/is easy category, which more or less translated into easy and quick to do, but impressive to the untrained eye.

As you think about when best to exit your technology venture, use a slightly modified version of the Hard/Easy Matrix:

Hard/Easy Matrix Applied
to High-Tech Startup Exit Times

	Looks Easy	Looks Hard
Is Easy	Exit postprofitability	Exit preprofitability
Is Hard	Don't even try!	Exit prerevenue

- **Looks easy/is easy.** You have to see the technology through all the way into commercialization. If you don't commercialize the technology, someone else will. In other words, the barrier of entry for a competitor is low. An acquirer won't be willing to pay for a business that doesn't have any true commercial value.

- **Looks hard/is easy.** From the point of view of the potential acquirer, you have solved a hard problem. Yet, the technical implementation was relatively straightforward and not too costly. The acquirer gives you more credit for your accomplishments than you deserve. Consider selling the venture preprofitability.

- **Looks hard/is hard.** You are in a good position to spin a story around a proof of concept. From the point of view of the acquirer, you have solved a difficult problem, and the barrier of entry for a competitor is high. You have a time-to-market advantage, even if you haven't implemented a commercialization strategy yet. Consider selling the venture prerevenue.

- **Looks easy/is hard.** Stay away from the technology in the first place. If your nerdy self wants you to demonstrate that it can be done anyhow, don't expect much financial reward or admiration.

When you do decide to sell, because the right moment has come, act swiftly. The market and the company's outlook can change very quickly, and your window of opportunity may close before you know it. Acquirers have a tendency to drag out the M&A process, mostly because waiting reduces their risk. You have to push back hard because their risk mitigation means greater risk for you and your venture.

NEVER, EVER GIVE UP!

Lesson 1: If a founder or key employee decides to leave the company during an M&A transaction, you have a problem. Make sure you have a good story at hand to explain the situation and minimize the negative impact.

Lesson 2: If your investors force the company to exit, you can leverage the situation into a good deal for employees and founders.

(continues)

Since the transaction will greatly benefit from the cooperation of the employees and executives, the investors and the acquirer should accommodate the interests of the employee base.

Lesson 3: Think about good times to sell your company, and take advantage of the M&A opportunities when they present themselves. Good times to sell are few and far between, but they can happen preproduct, prerevenue, or preprofitability.

Lesson 4: When you decide to sell, time the event well relative to the metrics that make future commercial success plausible. For example, sell at a time of increasing revenue, even if the baseline is small.

Exit Strategies

Selling a highly profitable business is fun, and so is selling a high-growth business! But what do you do when you have been losing money ever since you can remember? When you are sitting on overwhelming investor preferences? When you owe the bank? Or when your top line hasn't changed appreciably over many quarters?

HIRING AN INVESTMENT BANKER

In the midst of startup difficulties, someone usually suggests hiring an investment banker. Whether the company is desperate to raise funds or to sell itself, the situation instantly looks a lot more manageable when responsibility can be handed off to someone who knows how to do these things. Investment bankers know the right people (those with money); they maintain and stay in touch with a pool of potential buyers; and they know how to squeeze the most out of a deal. The investment banker is the deus ex machina of the startup world. While this may be true on average, there is no guarantee that the hiring of a banker will be the solution to your specific startup conundrum.

A banker is supposed to make the M&A transaction happen by (a) bringing interested buyers to the table, (b) motivating them to make a bid, and (c) negotiating a favorable deal for the seller. Yet, when we finally sold ThingMagic, the only bidders had been companies we had worked with independently, and the final negotiation was concluded by our CEO. Did we have unrealistic expectations?

Our banker did make a major contribution. Each time Thing-Magic's M&A process stalled, he would step in and get it going again. The stumbling blocks were many, and they were mostly caused by differences of opinions among *our team members*. We had investors who wanted out so badly that it threatened to derail the process. We had founders who thought they were worth a lot more than the deal would pay and hated the idea of signing noncompete agreements. We had employees who needed to be taken care of to carry the company forward. These issues caused the negotiation to come to a screeching halt more than once. Each time, the banker would somehow manage to get everybody back to the table and talk.

The investment banker in an M&A negotiation has one objective only: getting to close! Like real estate brokers, bankers benefit proportionally to the size of the transaction, but the first and foremost objective is to get paid at all. Hence bankers are perfectly motivated to remove all of the obstacles that threaten a deal. Whatever it takes to get it done, the banker will do just that, and as long as everybody eventually signs the closing documents, the means are justified.

The investment banker is ultimately responsible to the stockholders of the company; not to you, not to the investors, and not to the management team. However, alliances can cut many ways. Old friendships may matter more than principles of governance. Also, your banker can find herself in a serious conflict of interest relative to the buyer. A good friend hired a top-tier investment banking firm to advertise and sell his very successful and profitable company. After a long bidding process, the high bidder happened to be working with the very same banking firm. Even though different teams within the firm represented the two sides of the deal, to this day, my friend has a bad feeling that the outcome would have been more favorable had there not been a conflict of interest on his banking team. After all, a banking firm won't get future business out of an acquired company, but it can continue to work with the acquirer.

A banker does not change the value of your company, but a prominent firm can add credibility to the process. Potential buyers may be more likely to have a look and make an offer. Just be aware that by the time you are receiving bids or are in the final negotiation, what counts are the hard facts regarding your business: the top line, bottom line, number of customers, and above all the value and promise of your technology.

When selling a technology startup, the less revenue you have, the more you need to focus on intangible assets, including and most important, intellectual property. The value and competitiveness of your technology and know-how needs to make up for everything you lack in financial performance. The banker, of course, does not know your technology as well as you do, nor is she trained to understand it quickly and represent it to a buyer. The banker may or may not be familiar with your industry and the different angles of interest among the potential bidders. How could she make a good case selling your technology and business without significant help from you?

As you work with the banker, try to assign responsibilities in such a way as to leverage the banker's and management's respective strengths. Let her and the lawyers do the *unpleasant negotiating* work that could possibly threaten your relationship with the buyer. Since you will have to work for and with the buyer's team after closing, you better make sure your relationship is healthy by the time you move over. Also, let the banker focus on the administrative aspects of the deal including these:

- Convincing various stakeholders that they are getting a fair deal

- Managing the due-diligence process

- Managing the flow of information in a bidding war

At the same time, you should maintain control of other important aspects of the negotiation including these:

- Presenting and defending the value of the company

- Negotiating the fate of employees and executives after the acquisition

- Maintaining a backup plan for the company in case a favorable deal cannot be reached

- Building a relationship with the acquirer for you and your employees

The banker ultimately does not care much about these secondary aspects of the deal. If you don't take charge, these issues will be neglected in the negotiation. Make sure to stay close to the negotiation, and don't for a moment think you can sit back and let the professional handle the process alone.

POOR (UNPROFITABLE), BUT PRETTY!

The chances that your startup is generating substantial profits by the time you are trying to sell it are unfortunately slim. Rather, most of us attempt to get significant economic return for ventures that continue to lose money, rely on ongoing investments for operations, struggle to pay the bills, and generally live on the edge of financial viability.

In the absence of stellar financial performance, you are forced to paint a pretty picture, spin a story, and seduce potential acquirers in other ways. You have to come up with credible answers to a set of impossibly difficult questions: Why would anyone spend money on an entity that has been trying to turn a corner for the longest time? Why would anyone buy into the value of a technology nobody is using yet? Why would anyone believe that *next year will be big*, when they know all too well that it was going to be *next year* ever since the company was founded?

As you put together your story, don't get carried away giving un-realistic forecasts and exaggerated prospects for the postmerger entity. Your acquirer will make sure to hold you accountable for the expectations you create. Any disconnect between pretransaction promises and posttransaction reality will be flagged and held against the team. Be enthusiastic, but realistic!

Strategic Value

When businesspeople feel the urge to go down a financially unjustifiable path, they typically argue that the activity will bring *strategic value*. A business development manager who is eager to go forward with a project but can't get the customer to pay will suggest doing it for "strategic reasons." A product manager who would like to develop a product but is unable to project sufficient revenue for it will claim the new offering is "strategically important."

If you are trying to motivate an acquisition with strategic value, spell out in detail what you mean. If the *strategic value* cannot be shown to result in future cash flows and financial results, what good is it? Of course, you can't predict the future, but you can anticipate and describe and model scenarios under which the combination of the two entities will make a strategic, that is, financial, difference down the road.

The more detailed you are, the more powerful your argument will be. Use the opportunity to show off your understanding of the acquirers' business. Point out why the acquisition is good and strategic for *their* business. Startup management teams have a natural tendency to talk a lot about the value of their company. That is what they know best, and that is what they believe they need to promote. Yet, the important question for the acquirers is how the acquisition can improve their business, not yours.

Much has been written about improving sales success by asking pointed questions and putting oneself in the shoes of the customer.[1] Selling your company can benefit from the same idea. The acquirers need to see clearly that the acquisition will improve the value of their company. As the *salesperson* for your startup, you need to explain why that will be the case, much as a good product salesperson does the thinking for her customer.

Positive Trends

Startup valuations thrive on first and second derivatives! If the young enterprise can show a positive trend (that is, a positive first derivative) on any one key metric, it is in a good position to negotiate a favorable deal. Even though the current performance suggests a zero or negative company value, acquirers can be made to believe that the positive trend will continue and eventually result in profitability.

Even better, a positive second-order derivative on a metric suggests higher-order growth or exponential growth: now the entrepreneur can argue that the positive trend follows the infamous hockey-stick pattern and that the accelerating growth will fundamentally change the financial picture for the startup or its acquirers (Figure 12.1).

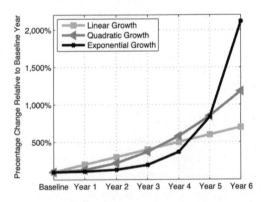

Figure 12.1: Example for positive trends: linear growth (positive first derivative), quadratic growth (positive second derivative), and exponential growth (exponential first, second, and so on, derivatives). Note that the three curves are very similar in the first three years after the starting reference point. It is practically impossible to detect the difference between the growth patterns at that stage. Also note that in this example it takes four years for the exponential growth curve to overtake the linear and quadratic growth curves.

As you promote the value of your company, realize that it is easier to convince someone of a startup's bright future based on a small
number of positive metrics rather than on a large number of mediocre metrics. Paradoxically, flat revenue with exponential growth in
another dimension is worth more than a little bit of revenue growth
and nothing else positive to say. As you are getting ready to prepare
your company for an exit, focus your energy on a smallish number of
metrics. While you should always put revenue on a growth curve if
at all possible, it may be a lot easier to spin your story around other
dimensions of your business including the following:

- **Sales of development licenses (software business) or development kits (hardware business).** At ThingMagic we used to brag
  about the increasing volume of development kits under the notion that those kits would translate into significant OEM product sales soonish.*

- **Customer profiles.** One of ThingMagic's VPs of sales introduced the notion of a $50,000 club and demonstrated how
  the number of customers with revenue in excess of $50,000
  had increased dramatically over the course of a year. Only 12
  months later, that was no longer true, but we discovered that
  the $10,000 club was doing very well that year.

- **Number of nonpaying users.** You might argue that customers
  who don't pay for a service or a product aren't really customers.
  However, don't make that point when you are trying to look
  good to your future owners. Rather, argue how you will be con-
  verting nonpaying users to paying users.

- **Growth margin.** The *growth margin* of a product or business refers to the ratio of gross profit over sales price and thus is an
  important metric especially for hardware businesses. Product
  growth margin can be used to argue in favor of a positive and
  bright future, no matter what the situation is. If the growth

---

* . . . until we realized that the conversion rate of development kits into product sales was unfortunately rather bad.

margin is good—that is, you are manufacturing products at a small cost compared to its sales price—management argues that the company is positioned well for growth and profits when sales will be picking up. If the growth margin is bad—that is, product manufacturing costs you a large fraction of the sales price—management argues that the bottom line performance of the company will improve dramatically as soon as the manufacturing process can be optimized with the help of the new owners.

Cost Synergies and Operational Savings

Ever since mergers and acquisitions were invented, cost synergies and operational savings arguments have been used to justify them. Yet, often the expectation that the combined entities will save money after the merger is flawed. Acquisitions that result in a new and independent division are more likely to increase the overall operational cost than reduce it (Chapter 13).

True operational synergies are impossible to achieve without reducing the personnel structure of the combined entity. While it is conceivable that the new parent will shift personnel from the acquisition into open positions at the parent, more likely, employees will have to leave to realize operational gains.

If you choose to sell your company based on the notion of significant cost savings after the acquisition, make sure you are willing to accept the consequences. Be prepared that not everybody in your organization will be working for the company after the transaction closes.

Market Reach

Smallish high-tech companies are notoriously limited in their ability to define, reach, and sell to their customer base. There simply aren't enough work hours and resources to effectively communicate with a worldwide market. If you spend your resources building up a channel, you will likely neglect the end users. If you decide to focus on the end users, you will likely neglect to build a channel. If you happen to balance these two requirements well, you still haven't spent a dollar on marketing and advertising. Also, while you are focusing on your domestic market, you probably will completely miss out on international opportunities, or vice versa.

In selling your company to a larger entity with established worldwide distribution, you have an opportunity to build on your new parent's distribution strength. You will have to figure out precisely how the merger of the two entities will provide the desired market reach for your venture. Whether you insist on going through this exercise before the acquisition or leave it to the posttransaction honeymoon depends on the circumstances. Perhaps you'll come up with a plausible story to promote the merger, and then adjust it to reality later.

(Complementary) Technology

Technology startups have no choice but to focus on a narrow slice of a technology, a solution, or a marketplace. Balancing the need to focus with the need to reach a market large enough to support the company is a terribly difficult task. Startups fail when they set themselves up for too small of a piece of the value chain or when they take too big a bite and choke. In the first case, the company ends up with a great invention and product but too little demand. In the second case, the company tries to do too much and cannot muster the resources to do it all.

An acquisition can cure either malaise, if done right. If the acquirer is in the business of providing solutions, the solution can absorb and leverage the benefits of the startup's technology. The acquired startup no longer needs to build and maintain its own channel and market. Rather, the team can focus on further improving the offering.

Alternatively, if the acquirer sells similar technology in a substantially overlapping marketplace, the combined entity can benefit from a joint approach in engineering, marketing, or distribution. Combining forces does not mean that people get laid off. Rather, it means that the combined employee pool can offer better coverage for all the little things that need to happen to make a product successful.

As you talk to potential acquirers, ask for details regarding their specific plans for consolidation after the merger. In addition, you should proactively demonstrate competency in outlining the opportunities that will arise from combining the portfolios and business activities. The more you learn about the acquirers' business beforehand, the better you can spin your story about opportunities and success for the merged parties.

You Are Missing the Boat!

Startup management teams like to think that their company is unique and one of a small number of worthwhile acquisition targets. If you wait too long, they tell the potential acquirers, the competition will have snapped up the few market and technology leaders, and you and all those who didn't move to acquire early will be left with unfortunate alternatives: (a) get into the market too late with a home-cooked sub-state-of-the-art product; (b) pay much more for a lesser startup asset; or (c) be left out of the new marketplace altogether.

Indeed, the number of worthwhile acquisition candidates in a high-tech field is usually limited. If you can manage to be among the top few vendors in your field, try the you-are-missing-the-boat argument. It is one of those versatile arguments that works particularly well with acquisitive companies who are not deeply familiar with your industry but fearful of losing out on the next big thing.

"You are missing the boat" is a good line to get a foot in the door with an acquirer. However, by the time you are entering a serious negotiation, you will need to back up your claim with more substantial arguments.

FINANCIAL VALUATION

The market for unprofitable high-tech startups is terribly inefficient. On the supply side, startups in a particular industry tend to be one of a kind for all practical purposes. On the demand side, the group of potential buyers is also small. Furthermore, information about deals, acquisition candidates, and interested buyers is impossibly difficult to obtain.

In the midst of talking about strategy and intangible benefits, it is easy to forget that the acquirer will be evaluating your company based on financial considerations. Even an obscure startup will be judged on whether or not some money can be made by it or with it in the future. While you may not like the picture that financial evaluation methods paint, you better be aware of what the formulas say. At a minimum, mastering the math might impress the MBA types on the acquiring team.

Comparables

Small technology companies are notoriously difficult to value based on financial performance. Hence, finding a comparable company with a known market value goes a long way toward establishing a baseline value. Since a one-to-one match is more or less impossible to find, you can use the gaps and differences to shape your story. If your *comp* has similar revenue but a better bottom line than your company, focus on the revenue comparison. If the comp has similar market share but much better operations and bottom line performance, argue that operations can always be optimized:

- **Public companies** with similar business models lend themselves as comparables (*public comps*). Since financials are publicly available, it is easy to establish metrics for valuation such as market capitalization over revenue, market capitalization over profit, or market share.

- **Recent acquisitions** of startups in the industry can be used to establish a baseline valuation (*transaction comps*). Financial data is not always available, and in some cases that is a good thing. The mere fact that a similar company was acquired suggests that there is value and interest. You can usually guess some of the parameters and emphasize the data that favors your own agenda.

- **Recent IPOs** offer the most favorable comparison opportunity. As much as possible, connect your own enterprise and enterprise value to such an event.* The S-1 filings of IPO candidates and the subsequent quarterly reports are the most comprehensive source of information for comparison you can hope for.

Enterprise Value: Pick Your Favorite Performance Metric

Comparables are best used in combination with the *enterprise value* (EV) valuation method, one of the simplest of the financial valuation methods. The EV is based on the assumption that the value of a company can be computed by multiplying a key performance metric with a fixed multiplier. The most common metric used is EBITDA or earnings before interest taxes, depreciation, and amortization:

$$V \quad = \quad EBITDA \cdot M$$

$$\text{where}$$

$$EBITDA \quad = \quad \text{current year's EBITDA}$$
$$M \quad = \quad \text{earnings multiplier}$$

The multiplier M is determined based on specific industry and market comparables and transactions. For example, high-growth industries tend to have higher multipliers than low-growth industries. Aside from EBITDA, other earnings metrics are being used depending on the situation, including straight *earnings* (E) and *earnings before interest and taxes* (EBIT).

Unfortunately, the *EV* calculation based on earnings metrics does not work for money-losing ventures. Instead, acquirers have adopted

---

* One of our competitors in the RFID industry went as far as comparing itself to an IPO candidate by their respective trade show booth sizes. The concept was interesting, but it ultimately failed because the comp wasn't able to go public after all. On the other hand, no harm was done. Be creative!

alternative metrics for companies that are money-losing, that are growing rapidly, or that are looking to sell for an outrageous amount in the midst of a valuation bubble. The metrics and multipliers can be very specific to the type of business or the market reality at the time of the valuation:

- **Sales: It is only a matter of time before our revenue will be overtaking our expenses.** *Annual sales* is the most widely adopted alternative metric in the EV calculation of nonprofitable growth companies. The multiplier applied to revenue and forward-looking revenue varies greatly with context and industry. It can range in value from less than one to multiple hundreds.

- **Number of Users: We have *chosen* not to charge for our services initially, but we will make lots of money when we do.** The *number of users* metric has made the dot-com boom possible and has recently been the basis for outrageously optimistic M&A valuations in social media. The metric works well for businesses where the cost of acquiring and maintaining a user is low.

- **Quarterly User Acquisition: At the rate at which we are going, we will have a large user base in no time.** If you don't have a lot of customers or users yet, show some numbers that explain how fast you are acquiring them. You will be credible if you can show a steady curve over many quarters or a sharp increase over a small number of quarters.

Discounted Cash Flow

The *discounted cash flow* (DCF) *method* of company valuation calculates the net present value (NPV) of an enterprise based on the assumed future cash flow (CF), and a discount or interest rate:

$$DCF \quad = \quad CF_0 \cdot \sum_{n=1}^{N} \left(\frac{(1+g)}{(1+r)} \right)^n$$

where

CF_0	$=$	positive cash flow in year 0
g	$=$	annual growth rate
r	$=$	annual discount rate
N	$=$	time horizon

The formula is simplistic in that it assumes a fixed growth rate and positive cash flow to start with. However, it is straightforward to specify cash flow explicitly in any individual future year including years of negative cash flow. Assuming you can rationalize sufficient positive cash flow at some point in the future, the formula translates the forecast into a positive DCF value. This, of course, is a welcome feature if you are promoting a startup for sale that has not yet reached breakeven. The more flexible formula looks like this:

$$DCF \quad = \quad \sum_{n=1}^{N} CF_n \left(\frac{1}{(1+r)} \right)^n$$

where

$CF_n \quad = \quad$ cash flow in year n

Let's consider the hypothetical example of the startup JuCo Inc. JuCo never quite made it to profitability. In the current year, the startup is expected to lose $500,000. The team has convinced the acquirer that the venture will break even the following year, show a profit of $500,000 in year 2, and then grow profits by 50 percent each year thereafter.

Using a discount rate of 10 percent and a 10-year horizon, we compute the DCF value of the company to be $10.8 million. Not so bad for a little company that continues to lose money!

Clearly the risk for a buyer increases dramatically as you push any assumptions about positive cash flow into the future. The challenge is to make a buyer believe the projections and then apply the formula. If you can manage to be credible about your financial prospects, the formal valuation method can help you get through the rest of the argument and land a better sales price.

Excuses

What do you do if all that math magic doesn't work?

Here are a few arguments you can make to the potential acquirers. Use them wisely and selectively! They have all been used before and only work in moderation:

- **"We are doing badly, but better than everybody else!** Our past and current financials are not the result of the management team's ineptitude or a lack of competitiveness. Rather, the lackluster financial performance is caused by unfortunate external factors, which are about to improve. We are the best team to bet on because we are weathering a bad environment well."*

- **"We are the market leader!** In the absence of strong financial results from any player in our industry, why not align yourselves with the leading startup? While we have not been able to capitalize on our competitive advantage quite yet, we are in the best position to dominate this marketplace, once conditions improve."†

- **"The market conditions are improving rapidly!** You should, of course, make your own assessment of our market. It hasn't been great up until very recently, but the signs of a major uptick in activity are definitely there. Just look at how company X [a customer] is initiating projects implementing our technology. Or look at how company Y [a competing vendor] is ramping up staffing and capacity to address the increase in demand."‡

---

* Hopefully, you are the best team in town. If you are not, build your argument on something you know how to do well.

† No need to be specific about timing.

‡ Use microeconomic arguments to suggest a changing macroeconomic situation in your market.

- **"If you acquire us, we will get much better access to manufacturing services, and we will improve our gross margins dramatically.** We are currently not making much money, but that is only due to the fact that it costs us too much to make our product. Once we are part of your organization, improved gross margins will enable profitability and more room to encourage growth through better pricing."*

- **"Recent product releases will increase our market share multifold.** The recent product we developed is a true quantum leap for our industry. There is nothing like it out there. Of course, the adoption process has taken some time, and we have to be patient. However, the timing is perfect. If you buy us now, you will benefit from the significant and imminent revenue increase. You just have to move quickly!"†

Bidding Wars and Skirmishes

You want to make it known to bidders that there are other interested parties in almost all scenarios. Nondisclosure agreements (NDAs) likely prevent you from disclosing any hard information, but you can be as suggestive as you want. The inability to disclose names and facts can work in your favor when the bidders are actually not in a relationship that would stimulate competitive anxiety.

Be careful not to mention other bidders inadvertently. In a tense negotiating situation in which you correspond and talk to multiple parties at the same time, it is easy to get them mixed up and mention the wrong company name to the wrong negotiating partners.

Large corporations and their decision makers are not above emotion when it comes to acquisitions. The desire to own a startup asset can

---

* Suggesting that both growth margin and pricing will change to the positive is somewhat of a circular argument. You might as well go all out and flatter them as best you can!

† Comparison between products is a highly subjective undertaking. You hold all the cards because you know the competitive landscape better than anyone. However, try to find an *independent* third-party reference to say something nice about your technology.

easily trump rational evaluation of an M&A proposal. You are best positioned to exploit the potential acquirers' emotional vulnerabilities if you find yourself a pair of interested buyers who have been archenemies or rivals for years. The two candidates should have enjoyed a history of beating each other to the punch or making each other look bad in past acquisition wars. Nobody likes to lose a second time to the same opponent, most certainly not a high-strung, alpha CEO. These potential acquirers will do everything to win and take the trophy acquisition home, rather than conceding defeat to their archenemy.

If you don't have an outright bidding war, at least try to keep a few interested parties lined up. At the various stages of negotiation, alternative options help keep the discussion with the leading bidder focused, and they provide fallback options when a deal falls apart. Carefully choose the moment when you mention secondary bidders. It is best to slowly build up tension going from the *expressed interest stage* to the submission of the *best and final term sheet*. This establishes credibility, whereas the suggestion of a second bidder at the last minute makes it blatantly obvious that all you want is a better price.

PROTECTING THE TEAM

After many months of advertising ThingMagic for sale, we finally had two offers on the table to buy the company. Alas, the consideration from the sale would not cover the outstanding debt and satisfy the expectations of the preferred stockholders.

What followed was a painful negotiation between investors, the management team, and the potential buyers to work out a deal acceptable to all parties. The investors needed as much money as possible out of the transaction. The executives and employees needed to be compensated for the sacrifice of making a multiyear commitment to the acquiring company, along with uncomfortable noncompete agreements. The potential buyers needed the commitment of the staff to take the business and the technology forward.

As a management team, we were determined to bring our employees with us and make sure that they were adequately compensated.

To support the need to incentivize employees, I offered our investors the following metaphor: "I feel like I'm being talked into divorcing my wife of 10 years, even though I still love her. At the same time, I'm being asked to marry a new woman whom I do not really know and who hasn't been particularly quick to express her love. Why would I agree to such a move?" Needless to say, neither the investors nor the potential new employers were impressed with my poetry.

Even though the emotions went out of control many times, we ultimately worked out a compromise that did achieve all the essential objectives: (a) sell the company, (b) return money to the investors, and (c) keep the employee base intact. In fact, three years after the deal closed, the majority of our people were still working for the new employer.

The more difficult the startup's situation, the more important is the retention of the core team in the M&A negotiation. Conventional wisdom suggests that retention of key personnel is primarily in the interest of the acquirers, while other stakeholders shouldn't care much. In reality, the commitment of the core team is in the interest of everybody involved. Any sophisticated buyer will simply not complete the transaction unless the commitment of key employees is reasonably assured. In a situation in which the venture has few other options than selling out, it is imperative to come up with a structure that keeps the team on board for the foreseeable future.

Employee Carve-Outs: Getting the Team over the Line

Most instruments designed to retain and motivate management and employees are sponsored by the acquiring company. The exceptions are *carve-outs*. Carve-out arrangements reserve a portion of the proceeds from the sale of the company for the benefit of the management team and the employees. The arrangements are made to motivate employees to work toward a lucrative exit for the company, when more conventional instruments such as stock awards or options wouldn't be effective.

Carve-outs are predominantly granted in M&A situations in which the startup's common stock is worthless. If the company is expected to sell in the preferences, no amount of common stock awarded can change the reality that the employees will not get a penny for their equity. And yet a successful deal depends on the co-operation and hard work of those very employees, leading up to the transaction and possibly beyond. When the board recognizes that the usual incentives are unsuitable to motivate the management team and employees, they may decide to allocate a cash-based employee package to be paid out when the M&A transaction closes.

Carve-outs can be instituted significantly in advance of an M&A transaction or during the final negotiations. When the decision is made to sell the company and it is apparent that common stockholders will have little or no financial upside from the transaction, employees tend to concern themselves with finding new employment rather than dedicating themselves to their current job. The carve-out arrangement made in advance of the transaction should help motivate the employees to stay put and hold off on career decisions until after the transaction.

Mind you, carve-outs do not constitute severance arrangements. Rather, carve-out packages are designed to motivate key employees to achieve a very specific goal for the company. Its purpose is specifically to NOT sever employees but, instead, to keep them in place and aligned with the interests of stockholders and the board.

Retention Bonuses and Earn-Outs

How do you put a price on people in an M&A situation? Can a buyer really hope to get value out of a management team that worked for a small independent startup for a reason? Unfortunately, nobody has found the perfect solution that would assure the cooperation, commitment, and passion of the ex-entrepreneur and her team.

Acquirers have the best chance of retaining key personnel if they offer a clever and tailored combination of financial rewards and *soft values*. The former entrepreneur or startup employee continues to look for personal fulfillment. She is still the person with little interest in a nine-to-five corporate job. Hence, the new employer needs to provide an environment in which an entrepreneurial attitude is appreciated and creativity is encouraged. In addition, the acquirer needs to offer reasonable financial incentives that take into account the personal situation of the individuals.

In the heat of the M&A process, it is easy to overlook the differences among various retention instruments. Most important, you should be aware of the differences between incentives to reward retention and those designed to reward performance.

Retention bonuses are paid provided that the employee stays employed for a well-defined number of months or years. As long as you don't give your new employer a reason to fire you for cause, all you have to do is stay put and cash in. Retention bonuses can be paid in cash or in stock awards, and they typically vest over the retention period. If you are given stock, factor in the possibility that the stock price will go up or down over the course of the vesting period. During the earn-out period, you will be overexposed to your employer's stock price without the ability to sell.

Performance bonuses, in contrast, are paid out if the individual continues to be employed and meets certain performance criteria. The acquirer likes to hold the management team of the acquired startup accountable for revenue and profit estimates by tying their compensation to those very metrics. The performance criteria may be based on the individuals' performance, the performance of the division, or the performance of the corporation overall. It's a nice hedge for the acquiring corporation, but it is potentially a very bad deal for the individual employees and the stockholders of the acquired startup. As you negotiate for a retention package as opposed to a performance bonus, consider the following arguments:

- **There's no absolute accountability without absolute control.** The management team will not be in control of its destiny after the acquisition. The structure that you are thrown into can change any day, and along with that the outlook for the business. Senior management at the parent company may kill any effort or significantly change the direction in their sole discretion. That is the natural right they acquire when they buy your company. Hence, the management team's compensation should not be a function of the venture's performance after the acquisition.

- **Feel free to have another look before you buy.** The acquirer is making a bet on the company after going through a long due-diligence process. The acquiring corporation has had sufficient opportunity to convince itself that the deal is a favorable one. Whether it ultimately is depends on many factors, almost all of which are out of the hands of the original management team. An acquirer's decision-making and due-diligence team likes to hedge the risk of the acquisition by offsetting some cost against the future performance of the unit. While this approach is understandable, it is not something the management team of the startup should agree to.

- **Future upside and future risk are changing hands.** The whole idea of a change of control is based on the premise that control, risk, and future upside change hands. The acquirer buys the future potential of the company, in addition to the current value. The previous owners including the management team used to own that upside. After the transaction they do not. Diluting the sales transaction with ongoing incentives tied to performance is only justified as an additional form of compensation for the management team, not as the primary retention mechanism.

In summary, retention packages recognize that the net present value of the startup requires the presence of the founders and management team going forward. These packages make a lot of sense for both the employees being acquired and for the acquirer. By contrast, performance-based packages can easily turn out to be disastrous for employees and management.

How Long a Commitment Should I Agree To?

Depending on what the acquirer wants to do with the new subsidiary, the requirements for key people to stay will vary greatly. If the startup has matured to the point where no heroic efforts in the day-to-day business are required anymore, the acquirer may actually prefer to hire cheaper replacements or put some of its own employees in place. If, on the other hand, the startup is still searching for a scalable business model or struggling to reach profitability, the acquirer wouldn't want to let the old team off the hook.

For me, a multi-year commitment sounded like a very long time. On the other hand, after 10 years of running a startup, I had no illusions, and I knew that a few years pass quickly if you are trying to accomplish something of lasting value. Two or three years is actually a rather short period to do something significant in a new environment. It is also just barely enough time to learn the ins and outs of life within a large corporation. I did appreciate the opportunity to experience that *other world*. The earn-out years allowed me to think about what to do next in the relative comfort of a secure job with perks designed to keep me interested.

In hardly any M&A situation can the management team get away with less than a year's commitment. Three or four years are within reason but certainly on the long end of the spectrum. As you negotiate the time period of your commitment, don't forget to pay close attention to other key terms in your new employment contract. Terms to look out for are the vesting of your retention stock or options, the noncompete provisions and their enforceability, and the severance provisions in case of a premature termination. Also try to anticipate what would happen if your new employer changed ownership during your committed tenure there.

ALL'S WELL THAT ENDS WELL

Lesson 1: If you intend to sell your company, an investment banker can help find interested parties, manage the negotiation process, resolve interpersonal issues, and get things moving again when the M&A process stalls. However, you yourself have to take a proactive role in selling a buyer on the value of your technology and company.

Lesson 2: If you want to sell your company at a decent price, make sure you can show growth along some aspects of your business. A small number of growth metrics is more helpful than mediocre numbers in all dimensions of the venture.

Lesson 3: Determining the value of an unprofitable technology company is a crap shoot. Be bullish about your business and industry, but do not overpromise! Then use some formulas to add credibility to your projections.

Lesson 4: The worst possible nightmare for an investor who wants to sell: no deal and a management team jumping ship. Don't be shy about asking for employee retention bonuses. Retaining key talent through an M&A process must be a joint effort of seller, buyer, and management team. If the best people decide to leave, everybody loses.

CHAPTER 13

Corporate Life

Life is full of misery, loneliness, and suffering—
and it's all over much too soon.
—Woody Allen (1935–)

Entrepreneurs are known to loathe the earn-out years following the acquisition of their company. So much so that some of us forfeit a significant financial upside just to escape from this modern form of "indentured servitude" as quickly as possible.

Not I! I woke up the day after the acquisition with a strange sensation of calm. I was going to have job security for a few years, and I didn't really need to think about the future for a while. What a relief after a decade of worrying! There would be enough opportunity to worry again when founding my next startup . . . but then, surprisingly, the worrying commenced that very day, working for a billion-dollar public company.

ACQUISITION PHILOSOPHIES

The companies that had expressed serious interest in buying Thing-Magic offered very divergent philosophies and plans for the business after the acquisition.

The most extreme bidder was interested exclusively in our patent portfolio. The new owner would have dismantled our company altogether and only kept the patents. Not a very flattering offer, and one we were glad to pass on.

The second bidder was interested primarily in the technology base and some of our engineering staff. ThingMagic as an entity would have ceased to exist, and our technology would have been integrated with a larger business unit of the acquirer. The majority of our employees would have lost their jobs, no matter what the official line was during the negotiation. In addition to the personal hardship for our employees, we expected the ThingMagic brand and legacy to disappear.

The third bidder and ultimate winner entered the race with the opposite proposal. While ThingMagic was going to lose its status as a separate legal entity, it was going to be preserved as an independent structure with its own P&L, brand, personnel structure, and management team. Everybody was going to keep their jobs. ThingMagic, the division of a multinational corporation, looked and felt very much like ThingMagic, Inc. Some long-term customers assured us that they didn't even notice anything had happened.

Divisional Acquisition

Post-acquisition life as a self-contained division certainly has a lot of appeal. The former startup lives on, and the employees are allowed to continue doing what they used to do. However, divisional acquisitions have significant downsides as well:

- **Did anyone mention *profitability*?** From the point of view of the parent corporation, a divisional structure is straightforward to manage, and it requires minimal corporate oversight. The general manager has the well-defined responsibility to run the business. Simple financial metrics are the primary management and communication tools between the parent and the new subsidiary. In exchange for giving the divisions the freedom to manage themselves, the corporation, of course, expects the divisions to deliver profits and ultimately growth.

- **No synergies, but lots of corporate overhead.** In divisional acquisitions the majority of the employee pool is preserved to run the business. As a consequence, the new division does not enjoy significant cost savings. At the same time, corporate overhead burdens the P&L of the division. Consequently, the bottom line of the acquisition is more likely to deteriorate than improve post-acquisition.

- **The quarterly earnings announcement is always around the corner.** Public companies famously are on quarterly schedules. As much as CEOs like to claim that they are managing for the long term, the requirement to stand up in front of stockholders and present quarterly financial results takes priority over that long-term view. Especially if the corporate numbers are under pressure, there is no room to inject additional funds into an acquisition, even though the additional investment would be necessary to develop the full potential of the fledgling.

- **The other divisions are busy making their numbers too.** The new division is dependent on the rest of the corporation and sister divisions to get leverage for its products. Yet the sister divisions are trying to make their numbers too, while working on their respective roadmaps and development priorities. It is difficult to get help from colleagues who are under financial pressure themselves.

- **Why won't they pay me a bonus for a good forward-looking strategy?** Understandably, incentives in corporations are structured to award today's financial performance, not speculative future cash flow. It is hard for a large corporation to adjust incentive packages for the small entity they acquired, even if other metrics would make more sense and help the new division develop into something truly great.

In theory, the divisional acquisition of a technology startup should be the best structure to ensure that the technology will live and get the attention necessary to fully leverage its potential.[1] In reality, the constraints on the small division may cause the exact opposite to happen: without sufficient resources and additional investment, the new division has no degrees of freedom to fulfill its promise.

Full Integration

Alternatively, the acquirer may choose to integrate the acquired venture into an existing business unit. The integration approach offers many advantages but also significant drawbacks:

- **Great cost savings, but you may lose your job.** Certain job functions will need to be eliminated to reduce cost and improve efficiency. The redundant jobs typically include positions in manufacturing, fulfillment, general administration, HR, and accounting. A force reduction could also include positions in sales, support, and channel management.

- **Bye-bye, beloved products and customers.** Upon complete integration, your existing business may be sacrificed to the new plan and business model. That's a tough one to swallow for the startup team. After all, everybody poured their souls into those products. It may also be tough on your customers who believed in you and your offering and who now have to find a new supplier.

On the positive side:

- **Finally, there is a sales channel!** Even if you did okay as a startup, your sales and distribution channel was probably suffering from significant constraints. As you integrate with the acquirer, leveraging the existing sales and distribution channel may just be the low-hanging fruit that helps you boost your performance.

- **Integrated product groups create integrated products.** The use of your technology within another business unit is a lot more straightforward when you are part of that business unit. Within the integrated division, the product marketing people and engineers of both the parent and the acquisition can work closely together to produce new innovative solutions that are better and more valuable than the sum of their parts.

There is no perfect answer to the question of whether to integrate or to preserve the startup as a division. In reality, the acquirer may not be open to a discussion on the matter. Acquisitive corporations have their own style and M&A methodology. A holding company tends to acquire companies and run them as subsidiaries, while integrated technology companies tend to integrate.

LIVING UP TO ONE'S OWN PROMISES

"Post-acquisition, revenue will explode thanks to the capability of our team operating under the powerful brand of the new parent."

"As we leverage the parent company's distribution channel, our technology will finally reach the market it deserves."

"The market is about to take off, and this is the last time an asset like ours can be acquired for an affordable price."

The pre-acquisition sales pitch is full of promises made with conviction and meant to accomplish a very specific goal: sell the company! Unfortunately, much as the salesperson is held to his forecast and target, the startup team is expected to make the vision that was advertised during the M&A negotiation a reality.

This expectation, of course, puts you right back in the situation you tried to escape by selling! You have spent a few years to a few decades trying to make a business work. You may or may not have reached profitability, but you were successful enough to sell your company. Then you realize that your new bosses expect you to make money and deliver healthy, double-digit operating margins every single quarter, much as your investors and stockholders expected prior to the merger.

The good news is the sale is done, and you are sitting on the same side of the table as your new bosses. The bad news is that all the issues you were fighting before the acquisition are likely still there. Being part of a larger corporation isn't in itself going to solve any fundamental problems for you. On the other hand, the acquisition is definitely a good time to initiate drastic changes to your business model.

Most important, make sure your technology or product is used for a purpose. Technology is a means to an end, not an end in itself. Find your killer app within the larger corporation or in connection with your parent's other product offerings. The acquirer had some application in mind when they bought you. Make that product vision a reality.

If your market was too small before the acquisition to sustain you profitably, it likely still is. Actually, it's worse now. Your large parent is not interested in small markets even if the market is able to sustain your little division. Therefore, get out of a disappointing market as quickly as possible, or at least branch out in new directions.

If you previously lost sleep over your products being commoditized, the fear of Far East suppliers stealing your business will keep you awake after the acquisition as well. Your new parent may be able to get you better deals on your manufacturing efforts. However, that does not change the basic fact that it is cheaper to design and make products outside North America. Rather than trying to salvage a lost cause, your energy may be better spent developing more complex offerings with the parent company. Through the combination of the technology assets of parent and startup, new products and solutions become possible. Think about which ones are most likely to hold up against overseas competition.

Change, unfortunately, takes time, even if the conditions are perfect. As you try to leverage the new institutional situation, you will realize quickly that nothing happens instantly, certainly not within a large corporation. While the parent is focused on instant quarterly results, any positive financial impact stemming from the acquisition is likely many months, if not years, away:

- **Roadmap and planning.** The startup days when a small team decided what to develop based on a gut feeling are now over. Corporate product marketing has taken over, which means detailed planning, market studies, priority setting, and product portfolio analysis. Development of the roadmap takes just about as long as the actual product development. If your brilliant idea makes it past all the committees, it is probably last in line behind several other high-priority projects.

- **New product development.** However you twist and turn it, it takes a large company a few years to come up with a new and sophisticated product. Not that you used to cut corners before. You did what you had to do: hardware revisions, quality assurance testing, beta testing, regulatory certification, environmental testing. But boy, do all these tasks take time in a large company with lots of processes!

- **Sales channel and partner development.** Let's assume the merger allows you to leverage the sales team and partner network of your new parent. Everybody is on board to distribute your product in addition to the existing portfolio. You may have thought that revenue would start flowing on day one, but instead, you find yourself training the sales team, explaining the new offering to partners, adjusting fulfillment processes to handle the new SKUs, and managing a marketing campaign to promote the offering to end users.

- **Customer adoption.** Customers took their time to adopt your innovative products prior to the acquisition. They still do. If your new products make sense, someone will use them eventually. Just don't expect users to drop everything and buy your gadget because you now carry the name of your well-known parent.

While you are desperately trying to do all these things in the background, you get beaten up regularly for your financial performance. Your new bosses want you to perform well right now. As far as they are concerned, they have spent enough money on the acquisition itself, and they are in no mood to now fund the operation as well. In fact, while the purchase price itself was capitalized, ongoing losses from operations are hitting the P&L of the corporation every single quarter. A big nightmare for the CFO on the quarterly earnings call!

What to do to address this fundamental tension between short-term numbers and long-term impact?

- **Remind your new bosses of the reasons they bought you in the first place, and ask for time.** They should appreciate that the startup was acquired for its future potential, more so than the immediate financial impact. If they are not willing to invest a little extra time and resources to unleash the startup's true potential, they are being penny-wise and pound-foolish.

- **Designate a full-time team within your organization to initiate new programs that leverage the parent company.** Assign a member of your marketing and sales team to explore and implement new channels. Task a product manager with the development of the new product offerings. As you designate these *ambassadors of change*, try to put them on their respective tasks full-time. If they are sharing their time between *old* and *new* initiatives, they risk being pulled in disproportionally by the legacy business.

- **Think of the new programs within the parent company as another startup effort.** The requirements on the team during postmerger integration are very similar to the experience of an early-stage venture. The tools for survival you developed back then will come in handy now. Whether you are trying to validate a market, find the resources to develop a product you are convinced will make a difference, or develop a customer base, you will face the same obstacles pursuing these goals now as you did when you started your venture in the first place.

———

Despite the difficulties you might face, try to enjoy the first few months following the acquisition. You'll get a short honeymoon, during which everything seems possible. You are the hero of your own people, and your new colleagues love to hear your story. The excitement will soon be replaced by a new day-to-day grind, not much different from the old one. Have fun while it lasts!

BEFORE YOU START ALL OVER AGAIN

Lesson 1: To make an acquired startup successful within the new parent, the former startup should maintain enough structural integrity to remain effective as an organization. At the same time, a certain amount of integration with the parent is necessary to enable synergies.

Lesson 2: Assemble a dedicated team to leverage the parent company, its resources, and its market access. Don't let day-to-day business and the need to produce short-term results take your focus away from improving long-term performance.

Lesson 3: Go about the integration effort as if you were managing another startup effort. Many of the tools you learned earlier in this book and during your startup time are applicable when it comes to navigating the difficulties, politics, and goals of the parent.

Lesson 4: Try to enjoy the new environment and the benefits of being employed by a large corporation! You can get back into the startup game soon enough.

Notes

CHAPTER 2

1. Edward B. Roberts and Charles E. Eesley, "Entrepreneurial Impact: The Role of MIT—An Updated Report," *Foundations and Trends in Entrepreneurship*, vol. 7, no. 1–2 (2011): 1–149; and Vivek Wadhwa, Richard Freeman, and Ben Rissing, *Education and Tech Entrepreneurship*, Kauffman Foundation Technical Report, Kansas City, MO, 2008.
2. Ibid.
3. Walter Isaacson, *Steve Jobs*, Simon & Schuster, New York, 2011, p. 105.
4. Robert W. Fairlie, *Kauffman Index of Entrepreneurial Activity, 1996–2012*, April 2013, http://www.kauffman.org.
5. Anthony K. Tjan, Richard J. Harrington, and Tsun-Yan Hsieh, *Heart, Smarts, Guts, and Luck*, Harvard Business Review Press, Boston, 2012, p. 21.
6. U.S. Census Bureau, Business Dynamics Statistics (BDS), http://www.census.gov/ces/dataproducts/bds/data_firm.html, 2005.

CHAPTER 3

1. David A. Vise and Mark Malseed, *The Google Story: For Google's 10th Birthday*, updated ed., Bantam Dell, New York, 2005, 2008, p. 45.
2. Uniform Trade Secrets Act with 1985 Amendments, http://www.uniformlaws.org/shared/docs/trade%20secrets/utsa_final_85.pdf.

CHAPTER 4

1. Robert W. Fairlie, *Kauffman Index of Entrepreneurial Activity, 1996–2012*, April 2013, http://www.kauffman.org.
2. Edward B. Roberts and Charles E. Eesley, "Entrepreneurial Impact: The Role of MIT—an Updated Report," *Foundations and Trends in Entrepreneurship*, vol. 7, no. 1–2 (2011): 1–149, p. 6.
3. MIT International Students Office, *Foreign Students at MIT, 1998–2011*, http://web.mit.edu/iso/about/statistics.shtml.
4. U.S. Citizen and Immigration Services, 2012, www.uscis.gov, and "Green Card Caps," H1 Base, Inc., St. Petersburg, FL, 2012, www.h1base.com.

CHAPTER 5

1. Seed-DB, 2013, http://www.seed-db.com/accelerators.
2. Y Combinator, 2013, http://ycombinator.com/atyc.html.
3. www.sbir.gov.
4. www.ssti.org.
5. Edward Berenson, *The Statue of Liberty: A Transatlantic Story*, Yale University Press, New Haven, CT, 2012, p. 86.
6. Kickstarter, www.kickstarter.com.

CHAPTER 6

1. Altman Weil, *Survey of Law Firm Economics, 2002 Edition*, Newtown Square, PA, 2002, http://www.altmanweil.com.

2. State of Delaware, Department of State, Division of Corporations, Dover, DE, 2012, http://www.corp.delaware.gov/aboutagency.shtml.
3. CNNMoney, "Fortune 500: Our Annual Ranking of American's Largest Corporations" for 2012 by state, http://money.cnn.com/magazines/fortune/fortune500/2012/states/GA.html.
4. American Institute of CPAs (AICPA), PCPS/TSCPA National Map Survey Commentary, 2008 and 2012, http://www.aicpa.org.
5. Cambridge Innovation Center, Cambridge, MA, http://www.cictr.com.

CHAPTER 7
1. U.S. Small Business Administration, "Insurance Requirements for Employers," Washington, DC, 2012, http://www.sba.gov/content/insurance-requirements-employers.

CHAPTER 8
1. Fenwick & West LLP, *Trends in Terms of Venture Financings in Silicon Valley, Q2 2004–Q4 2013*, http://www.fenwick.com/publications/pages/default.aspx.
2. Ibid.
3. Ibid.
4. Ibid.
5. Ibid.
6. Ibid.
7. Ibid.

CHAPTER 9
1. Mary Catherine O'Connor, "Alien Technology Closes $38 Million in Funding," *RFID Journal*, October 6, 2008, http://www.rfidjournal.com/articles/view?4363; CrunchBase, http://www.crunchbase.com/company/alien-technology; and NASDAQ.com, www.nasdaq.com. I should clarify that I have never been privy to nonpublic information regarding Alien Technology and its fundraising efforts. Rather, the information is publicly available from regulatory filings and the media.
2. Clayton Christensen, *The Innovator's Dilemma*, Harvard Business School Press, Boston, 1997.

CHAPTER 11
1. My cofounder Matthew Reynolds deserves the credit for coining the term *untested hypothesis* in this context. Eric Ries introduced systematic hypothesis testing for technology ventures in his book *The Lean Startup: How Today's Entrepreneurs Use Continuous Innovation to Create Radically Successful Businesses*, Crown Business, New York, 2011.

CHAPTER 12
1. Thomas Freese, *Secrets of Question Based Selling: How the Most Powerful Tool in Business Can Double Your Sales Results*, Sourcebooks, Naperville, IL, 2000.

CHAPTER 13
1. Clayton Christensen, *The Innovator's Dilemma*, Harvard Business School Press, Boston, 1997.

Index

About the Author

Bernd Schoner worked in the Physics and Media Group and the Things-That-Think research consortium at the MIT Media Lab, where he received his PhD. He cofounded and sold the high-tech startup ThingMagic to Trimble Navigation, a multibillion-dollar public technology company, where he now serves as vice president of business development. He lives with his wife and son in New York City and Cambridge, Massachusetts.